First Published August 2010

Revised October 2011

Howard Carr

THE LAST PEBBLE

The command that I am giving you today is not too difficult or beyond your reach. It is not up in the sky. Nor is it on the other side of the ocean. No, it is here with you. You know it, so now obey it.

Moshe Rabbenu circa 1500 BC

The people and politicians of the world, community by community, nation by nation, will now determine whether we can create and sustain the international vision, commitment and collaboration which will allow us to seize this historic special opportunity and to rise to the challenge of a planet in peril.

Nicholas Stern December 2009 AD

CONTENTS

A Century Defined — 11
The key issues for the twenty first century

What Question? — 17
Addressing the issues

The Unspoken Word — 25
The population issue

Heating Up — 43
The Climate Change issue

Every Crumb Counts — 79
Feeding the population

Liquid Life — 123
The fresh water issue

Transformation — 133
Delivering change

Not By Bread Alone — 163

PREFACE

My elder brother is a musician and artist. I continue to enjoy greatly his creative works and have learned from him that just playing the notes is not enough, and it is far better to put their delivery at risk in a search for timbre and musicality that makes a sheet of black notes into something that can transform a moment and transcend technical perfection. That has become my inspiration and source of courage to even try to write these pages.

The matter of mankind's present predicament is too big, has too many knowns and unknowns, threatens the gravest of potential outcomes and cannot be an easy read. Yet for several decades I have worked in and with the business of food from its start in the cold spring field to its display in a warm well lit supermarket. During that time I have tried to play each 'note' as perfectly as I am able; now I have taken the risk of transforming some of these notes into a single composition. This book is therefore not a comprehensive treatise, but rather a composition that I hope will play lightly on the ear and yet stir the heart at least a little; for all change is borne out of passion.

Howard Carr
Folkingham

A CENTURY DEFINED

There is a short scene early on in the spoof film Hot Shots where Dead Meat, who has everything in terms of skill as a pilot and intellectual capacity, kisses his loving and clearly devoted wife goodbye before a flight saying at the last minute, as he climbs into the cockpit, that he has discovered the secret to solving many global problems including global warming. His wife offers to take the paper, which will unlock such knowledge, but he brushes the offer aside and instead tucks it in a pocket of his flying suit saying there will be plenty of time for such matters when he returns. Of course in these films such happy men with everything the world can offer never return and the paper perishes with the hero and his crashed plane; but what was written on that piece of paper?

Like many people, I have for many years failed to understand how this earth could cope with an ever increasing population consuming necessities like food and water, and luxuries like cars and electronic goods. It defies logic and common sense. We have a finite sphere with finite resources be it land, water, or minerals. The only infinite resource is the sun and the light and heat energy it emits. It follows that with population growth continuing to surge and the greater industrialisation that it brings there can only be shortages. As members of that global population we, therefore, have a huge and life threatening challenge on our hands. This century will be defined by food and water supply and our ability to feed all or watch on as hunger and malnutrition sweeps the globe.

It would seem logical that the more sophisticated and complex the society the more vulnerable it is to such shortages. This is an inevitable truth, albeit initially it can be ameliorated by market forces, whereby money affords rising priced goods and in so doing denies the poor. It follows that initially the poor will suffer the greater hardship and indeed this is exactly what we are currently experiencing. However, the poor countries are largely agrarian economies and as shortages tighten then food for export will be denied and consumed internally to maintain stability. It is at this point that the wealthy importers will find that money is not enough. In extremis the only safe existence is that of the self-sufficient subsistence farmer. Welcome back to BC?

But we live in the third millennium of AD and whilst the questions around human existence have not changed between then and now the consequence of not finding an answer has moved from individual and group annihilation to the potential global collapse of the human species.

What is most disturbing about how these issues are discussed and represented is our tendency to use the future tense. Those days are long gone and we are now seated in the crisis itself. We only need to open our eyes and see the growth in industrialisation, the pull on commodities and the record prices it is engendering. Then there is the increase in carbon dioxide levels contributing to a rapidly warming world with all its well documented consequences as well as growing levels of hunger and malnutrition.

Staple world food stocks of grain sit at around 60 to 70 days with growing crops taking many days longer to mature to harvest. We therefore live off our stock, working hard to regenerate it and watching as that stock gradually shrinks under the weight of ever greater demand. Despite our best efforts there will always be times of plenty and shortage as agriculture interacts with the weather, and it follows that as our buffer stocks reduce so we will become increasingly vulnerable to such moments of shortage before the stock pile is finally exhausted and we can no longer feed ourselves.

I was fortune in my early career to meet such leading thinkers and writers as Walter James Northbourne and Fritz Schumacher. Lord Northbourne was the first person to identify and coin the phrase 'organic' in the context of farming. He was an intellectual, thinker, artist and writer who devoted much of his long life to agriculture. I was privileged to be given the opportunity to develop the vegetable production enterprise on his East Kent estate. He was an extraordinary man, distinguished, tall, well built but thin and of erect stature; a man, who having rowed in the 1921 Boat Race for Oxford, retained his athleticism of mind and body into his advanced years. He had an enormous impact on my life and thinking and I enjoyed his company, wit and the fine works of art he painted. Everything about him was quiet and possessed a certain reverence; he was a human cathedral.

Having revisited his writings I now realise what mixed feelings he must have had as he observed from his house the growth of much he abhorred on his very own land as the acres of intensively farmed vegetables and salads grew year on year. It was during these years that the tranquillity of the Garden of Kent with its numerous smallholdings was broken and transformed into large scale vegetable production; we were right at the vanguard of that transformation. He would have liked the fact that we employed 1000 people bussed in from Deal, Dover and the outlying villages to harvest and pack produce. That was a social contribution to the community, but the industrialisation of the growing using ever more mechanisation and chemicals in the process must have filled him with alarm.

During the fifteen years I knew him before he died, I came to share the board table, but never once did he voice conceptual dissent. The fact that he was a man of deep principle and integrity and yet lived with this conundrum only goes to underline just how deep the present dilemma for many of us is; we who are lesser men.

We like the advantages of travel, consumer choice and indulging in luxury and yet we equally know it has a price for ourselves and others and as a consequence we feel disempowered and unable to make the changes required. If Lord Northbourne had said 'enough' then the industrialisation of his farm would have ceased, but so would the jobs and income of more than a 1000 folk at the bottom of the earnings scale.

But these are the choices we have to make if as a generation we are not to fail those who would follow. The quantity of papers, books and articles analysing our current predicament is legion and much of it is of the highest standard. Inevitably each has its own angle of interest and emphasis, but the whole has become a multi-faceted jewel of analysis that shines clearly and brightly amidst the darkening clouds. The problem we face is how to respond and resolve, for like Lord Northbourne with all the best and well thought out intentions in the world, we are failing.

Failure is not an acceptable solution and these pages are a contribution to our understanding written by a practitioner in the worlds of farming and the food processing industry, who like someone in a smoke filled room desperately searching for the exit is sometimes random and sometimes logical, in panic and then calm, with muttered prayer and applied knowledge, always frenetic, but never giving up in the hope of survival.

But to find our exit we must first pose the right question.

WHAT QUESTION?

My grandfather was a self-made man. Born in London in 1883, he suffered under an alcoholic father, who boxed his ears until he became partially deaf. Somehow he found his way into the local Salvation Army, who gave him new hope and sound advice about living out the practicalities of life. I only really just remember him, but through the memories of my mother came to feel I knew him better than I could have done. He was undoubtedly a hard worker who looked for business opportunities and sometimes got it right and sometimes got is very wrong. Investments included the tin mining industry in Cornwall just before the tin price collapsed in 1920 and the opening of a leather goods shop in London just prior to the start of World War 1. Unfortunately the leather goods for sale were all sourced from Germany and this particular enterprise inevitably proved less than popular and soon failed. He did, however, go on to build a prosperous business in London's Highgate with two shops, a printing works and the ownership of the local newspaper. One of his sayings was 'bite off more than you can chew and chew it.' That piece of advice struck a chord with me and I have lived by it ever since. In seeking out the right question and then searching for its answer will require all his tenacity and determination.

Asking the right question is critically important if we are to find a way forward. Time is of the essence and we cannot afford the luxury of wallowing in endless hypotheses or engaging in academic work-outs. We need to bring focus to our dilemma and asking the right question will at least set us off in the right direction even if we flounder in finding its full answer. The importance of this was brought home to me in the early 1990s when I undertook some work for Sir Ian Morrow, who was by then coming towards the end of his active professional life. He was a renowned Scottish accountant and guru who was in effect an early, if not the first, company doctor. He was famously called into Rolls Royce in late 1970 as Deputy-Chairman when the company was struggling, having overspent on the development of the RB211 engine, which was being designed and developed to deliver more efficient thrust for Lockhead's TriStar aircraft. The company entered receivership in February 1971.

This was of immense importance not only to this country but also to America as Lockheed would also have gone into receivership as a result of Rolls Royce's failure; and they were key to the American military supplying them with the C-5A Cargo Plane and the Poseidon Missile. Consequently the UK's Conservative government under Edward Heath came under intense pressure from home, to protect jobs, skills and a British institution, as well as from President Nixon in the USA. Against their political instinct they decided to nationalise the company and Morrow was asked to become the Managing Director. His response was to roll up his sleeves, spend time on the shop floor, analyse everything and then work to a resolution. When he resigned in 1973 following a dispute with Michael Heseltine, the then Minster for Industry, the company was back on its feet and its survival assured. Ian Morrow was knighted and his reputation was well and truly enhanced.

The task that I was asked to undertake was of a rather more modest scale and involved a pharmaceutical company, of which he was chairman, that was led by a research scientist who had a deep conviction that lipid based drugs were often more efficacious than the more traditional amino acid styled ones. He had developed his business around a cohort of drugs based on fatty acids such as Gamma Linolenic Acid, which was sourced from Evening Primrose and Borage seed. My brief was linked to the supply side and commercialisation of products. This was not an easy assignment as people's reputations and thereby their credibility would inevitably be put on the line. At the outset Sir Ian's advice to me was to 'enquire, listen and smile' explaining that from his experience this would always 'elicit the truth'. This exchange has proved invaluable and in searching out the question we are to ask, this advice will undoubtedly prove as effective as ever.

We certainly need to be strictly analytical in our approach to defining the problem we face and the issues that lie at the heart of that problem and that are aggravating and exacerbating it. Once we understand the true global problem that lies ahead then we can enlarge our prognosis and begin to include the individual and his or her plight. Then the consequences of all that is going on are made real and immediate and it is then that our response must become emotional. For it is out of raw passion that change is liberated and can confront the conservatism of fear and the powerful protectionism of interest.

I am fortunate to know a newly qualified PhD student who has just completed her work on the rainforests in Central

Africa and the effects of various payment regimes and schemes to protect them for social and environmental reasons from the loggers, who year in and year out diminish their size. It became clear to me that the more she delved into the subject and began to understand the wider issues of climate change, human rights, justice and poverty the more she became depressed by what she saw and read. To her the future could only hold an inevitable collapse of society and an extremely bleak outcome for mankind. This is a hard burden to shoulder when you are in your twenties and it is the silent fear that must run through the minds of many of the young.

Interestingly her immediate response in searching for a solution was to seek refuge in ideologies that had been tried and found wanting during the 20th Century; they were all focused on political solutions and big government. We embarked on the debate with considerable vigour and discussed the driving power and failings of capitalism and communism and our inability to find the Third Way. For the age of communication my young friend seemed lightly educated in terms of the lessons to be learned from history and how the modern world actually functions. This caused me to be impatient, ham fisted in style and over derogatory in argument to the point where my frustrated friend claimed that 'I never set out a single new idea'. It was a fair criticism that struck home with some force and so I wrote her the following e-mail composed as I sat in the dentist waiting room.

I feel I owe it to you to explain in outline where I sit – remember I am not setting out the argument just the unsupported framework.

Let me say before I start that I am deeply concerned that people talk about the 'big crisis' as coming in 10, 20 30 years – we are in fact in it now and it can only get worse, so we need to act.

The first issue is one of human population. There are limits to growth in any bio-system. The key question is simply how many people can this planet support? Is it 1bn, 6bn, 9bn, or what?

That population number is not complicated – it depends on the following;

Food availability
Fresh Water availability
Commodities availability
Energy availability
Level of pollution

Each of these has to operate in a sustainable mode.

We are already experiencing food shortages – 1.1bn go to bed hungry or worse.
We are already experiencing fresh water shortages.
We are already experiencing shortage of some key commodities.
Energy is less of a problem in terms of supply – but a huge problem in terms of pollution. The energy supply issue can be technically resolved – the rate of transition to sustainable energy will inevitably be linked to the price of fossil fuels.
The level of pollution is impacting on climate already and is reducing food and fresh water availability.

The key question comes largely down to what level of population can be fed and watered sustainably?

We cannot defeat biology and the rules of nature – a small volcano in Iceland causes aviation in Europe to cease overnight.

Man is inventive, but even that invention cannot defeat nature's cycle of birth, death and re-birth.

It follows that as a doctor is sworn and committed to safeguard life, but nevertheless witnesses death so we must be committed to ease the pain of mankind. Acceptance of the biological cycle and its inevitability is not defeatism or failure, but rather an affirmation of who we are and our oneness with the eco-system in which we live. It is a prerequisite to our addressing the issues.

My hope is that we can use our considerable inventiveness and compassion in what must be a frenzy of activity to soften the 'curve' as population finds its sustainable level; as indeed it will and must if there is to be a future for mankind after we are long gone.

Take care

H

With all the sophistication of modern living with its science and technology our Question was grounded in a moment some 4 billion years ago when life took its first precarious steps on this volatile planet and began to evolve. In relatively recent history Darwin encapsulated that in his writings as he considered the conundrum of competiveness of survival and the interdependence of species;

'It is interesting to contemplate a tangled bank, clothed with many plants of many kinds, with birds singing on the

bushes, with various insects flitting about, and the worms crawling through damp earth, and to reflect that these elaborately constructed forms, so different from each other, and dependent upon each other in so complex a manner, have all been produced by laws acting around us.'

The question the world now faces is working with '(natural) laws acting around us' how it can feed the current population plus 2 billion plus extra mouths in a matter of just 30 or 40 years? That is a huge challenge. We have, however, faced and gained experience of dealing with such a quantum increase over the last few decades. However, as each billion of people is added to our global family the problem becomes increasingly difficult and the solutions to feeding ourselves more demanding. In truth in the short term we have little choice but to throw every resource of technical innovation and human ingenuity at the problem. The bigger question is how is this planet going to support such a level of human population on a sustainable basis? That requires not only technical and scientific innovation, but also rethinking as to life, its value and purpose. Welcome back to times BC after all, for the fundamental issues and questions have not changed.

Our question therefore emerges as being simply 'how many people can this planet sustainably feed?'

THE UNSPOKEN WORD

A matter of taboo

When I was in my twenties and between jobs I had the pleasure of driving my mother to Wokingham one early spring day to visit an elderly lady. She was Commissioner Catherine Bramwell-Booth, who lived at North Court in Finchampstead and was the grand-daughter of the founder of the Salvation Army, William Booth. Born in 1883 she had travelled England with William Booth at the age of just 20 as her remarkable grandfather, in his last years, preached his way up and down the country. Then as a serving officer she engaged in a lifetime of social work including trying to improve the lot of many in a Europe twice broken by the First and then the Second World Wars.

We arrived at her house, painted red and set in a rambling park-like garden, at the appointed hour and rang the bell, which clattered its way within. The door was answered and we were greeted by an elderly white haired lady in Salvation Army uniform who ushered us in to an extremely large drawing room, illuminated by natural light pouring in through the large windows on two sides, and containing several clusters of chairs and tables together with a clearly well used grand-piano. Catherine Bramwell-Booth was seated by the open coal fire, very erect and also in her uniform. She rose as we entered to welcome my mother, whom she knew well, and I was duly introduced. Her personality was quite electric and as she engaged in conversation it was immediately evident that she was a remarkable, precisely spoken, clear thinker, of immense character, who lived by conviction, with an almost unfathomable depth of caring, and driven and inspired by her family roots of which she was clearly very proud. Shortly after this encounter she was to appear on TV on the Parkinson late night chat show as his guest and she asked him the same question as she asked me – 'are you saved?' Her penetrating eyes and intense engagement demanding the truth for nothing less would do. A difficult question for a reserved young man, but I was consoled that Michael Parkinson with all his experience was to fair little better. She was in her mid 90's when I met her and she stood or sat tall and erect, her concerns and care for people still driving her with a real sense of urgency that was graced by a gentle, slightly mischievous and contagious sense of humour. Tea accompanied by Rich Tea biscuits was brought in on a white, starched serviette covered tray, and as we sat there I listened intently as she explained that she prayed everyday that she 'might be spared for a bit longer as there was so much to do'. The Commissioner, who had served the poorest of the poor, those afflicted by hopelessness and who had witnessed appalling wrongs and widespread destruction still burned with a sense of purpose and belief that nothing could stop the change and improvement in lives that her faith had and would bring. This unbounded enthusiasm for life and for the purpose it holds was to leave a deep impression on many including this young man. She was to die a few years later aged 104 bringing down the curtain on a remarkable life and leadership dynasty.

Catherine Bramwell-Booth's conviction and enthusiasm is exactly what this world requires as man now faces up to the greatest problem he has undoubtedly ever had to confront; over-population. It is interesting that any comment on human population is in danger of being treated as a no-go area. It is difficult for those of faith to question the over-existence of man for in so doing they perceive they are questioning God's creation and creative purpose. For the moralist and humanist there are issues surrounding the value of life and the choices it would bring as to who should live and who should not?

For all of humankind there is the innate biological mechanism of species survival that feeds on growth and adaption and demands greater propagation to enhance the chances of its survival. For the extreme poor of the world the question of survival has its own immediacy and has been brought to the individual level as part of the daily battle to live. This rich mix of morality, faith, intellect, basal instinct and enforced pragmatism, whilst all essential parts, nevertheless seems to make the subject of population and its control somewhat taboo.

Part of biology

The starting point has to be biology for we are animals fulfilling a role in a very complex ecosystem. Why we should deny such a role, as we so often seem so determined to do, is beyond reason for it suggests that we are somehow outside the animal kingdom, set apart and governed by different criteria and rules. Yet we grow, we reproduce, we expel waste, we breathe and respond to our environment. In other words we are biologically alive and perform the functions that define a living organism and of course, when critical cellular dysfunction occurs, we die.

Simply by taking in the beauty before us in field, garden, wood or park, or by looking down a microscope or peering upwards to the heavens cannot fail to engender a feeling of huge privilege and fulfilment for we are actually a part of this amazing system, creation or whatever is the preferred definition. In Darwin's day the fear of humans being linked to monkeys caused outrage and yet today in these enlightened times when we know that all mammals possess about three billion genetic base pairs and that the ape has a 95 percent level of gene similarity with us and even a mouse shares an 85 percent similarity, we will not let go of such a fear. We still harbour that wish to be separate and outside of our given home in the animal kingdom.

Only when we recognise our position and role within the ecosystem can we move on and address the question of population and how it is, and will, impact on our survival. Such a moment of realisation can only be liberating and enriching to our understanding of the life we live. Acceptance of the rules of biology suddenly lightens our burden and illuminates our path for we see that at its simplest level in response to plenteous food, water and a secure habitat then there is population growth. However, as that population grows it devours more food and water and despoils its secure habitat to the point where there is a moment of collapse and the population sharply falls away until it reaches a level of survival or extinction from whence of course it renews the cycle or makes way for other species. The more adaptable the organism the longer will be the lifespan of the species, and the more threatening the environment to survival the greater the innate urgency to multiply.

Man's ecological dominance has arguably come about as a result of his higher intelligence. This has enabled him to effectively harness the tools of nature he found scattered around him, be they flints, domestication of wild animals, use of minerals and fuels or pharmaceutical bearing plants. Such ability to modify the immediate onslaughts of the natural environment has undoubtedly influenced the shape of the population curve.

Initially as hunter gatherers the world carried a human population estimated at just a few million. Then approximately 30,000 years ago we started to farm and settlements became established and cities of urban dwellers evolved resulting in population growth to a level of circa 300 million. Now in more recent times industrialisation has opened up new resources, which with rapidly advancing technology has come to support our current population of 7 billion.

Draw the population count curve, and whilst we must accept that our ingenuity and intelligence has extended the time-scale dramatically, the biology of the curve remains in tact and as the human population races away so the point of its biologically predicted collapse comes closer. Indeed the current growth rate is already outpacing food and fresh water provision despite the effort of science, technologies and the exploitation of earth's finite resources. There can be no question that continued population growth will lead to even greater food and water shortages and despoliation of habitat; indeed it is already happening. We have been around now for 3 million years and the 'collapse' may take a period of 50, 100 or 200 years in duration; each is but a moment in our existence and fulfils the natural law under which we exist.

Population explosion

Today the population explosion is driven by birth rate and life extension. I was born into a world with a population of 2.3 billion people. In my lifetime that population has grown threefold and today the world holds 7 billion people. By 2050 it is estimated that with current uninterrupted growth this number will be circa 9.2 billion; an additional 33 percent. Of this projected growth over 90 percent will occur in Asia and Africa, particularly in the Subcontinent of India and sub-Saharan Africa, whilst in West European shrinkage will in part cushion growth in the Americas. This simple analysis highlights the problem focal points, which would have been even more acute without the one child policy introduced into China in 1978, which it is claimed still applies to an estimated 36 percent of their population.

The Chinese 'experiment' has brought to the fore a whole raft of controversies with increased female infanticide, forced abortions and of course hidden births. This policy is an anathema to many in the East and West, but it can claim that it has helped restrict population growth and it is argued that it still finds approval with three quarters of the Chinese population. But there are issues, not least sex selection be it by fair or foul means. In the European Union there are approximately 50 million males under the age of 20 years. If that whole generation was to be told that there were just 10 million girls for them to share their lives with then the social consequences would be unimaginable with an upsurge in violence, crime and dysfunctional men. In China that scenario has become a reality and today there are 40 million more boys and young men under 19 years of age than there are girls as a result of the manipulation of the sex ratio. Indeed this problem is endemic particularly in Asia with Northern India sharing with China a 100:120 girl to boy birth rate as a result of human interference caused by cultural prejudice, poverty, medical science or state imposed controls. Encouragingly South Korea seems to have resolved this issue from which they also once suffered through a mix of increased affluence, education and greater equality within society between the sexes. Even India now claims that 10% of women of child-bearing age have restricted themselves to a single child family; a family they can afford and educated well.

At the other end of life improved living conditions and health-care has extended life expectancy to a world average of just over 67 years and rising. Compared with earlier times of say the great civilisations of Rome or Greece or the less developed period of the English Middle Ages when life expectancy hovered around 30 years this increase has largely occurred in the 20th Century, which started with an estimated average global life expectancy of circa 30 to 40 years.

Conversely of course difficulties arise when populations fall away and the declining birth rate exceeds any increase in life expectancy; the population ages. In those countries experiencing such conditions it is estimated that the percentage of the population over 65 years of age will broadly double by 2050. This is the case for USA, UK and France with the percentage of over 65s rising to around the 50 to 60 percent level. More extreme cases are forecast for Germany where the predicted figure rises to 74 percent and for Spain and Japan it is set to exceed 90 percent. The financial and social pressures this level of 'non-productive' population will place on those of working age are yet to be fully addressed by governments. In Europe the ratio of four workers for each retired person is forecast to fall below two by 2050. This issue is another time bomb gently ticking away.

Change takes time

Population changes reflect the generational years and so with humankind any distortion in numbers takes time to resolve as the weakening waves eventually subside into the norm. I am a part of the massive baby boom that started towards the end of the Second World War, which then provided a second lesser wave in the 1980s, which the Americans refer to as the Echo Boom, and finally a third yet further reduced wave starting in the 'noughties' of the 21st Century. In other words it has taken 60 to 70 years or a lifetime for the UK population, excluding migration factors, to revert to its norm.

When the urgency of global matters is measured in units of 10, 30 or 50 years the process of stabilising or influencing population within that timescale can be seen to be infeasible. That is not to negate the need to address this issue for if we are to come through this 21st Century crisis and settle on a sustainable level of population then the work has to begin now. The question is whether man can mould his own demographically stable and sustainable global population or alternatively whether in the event of his being unable to do this he has to rely on nature's perhaps rather harsher remedies and shaping?

The debate

The debate over population is frequently high-jacked by those with extreme views and, whether out of good or ill intent, it becomes extremely emotional and often horrifying.

In modern history this debate gathered momentum in the later part of the 18th Century as newly discovered worlds in east and west began disgorging foods, fabrics and minerals into a Europe seemingly bubbling with ideas. Thomas Robert Malthus, whose writings since have been used and misused to support a whole plethora of arguments, engaged with the philosophers and thinkers of the French Enlightenment movement. In particular Nicholas de Condorcet, a nobleman living in the midst of the French Revolution, argued for widespread public education, a free economy, and equal rights for women and for all irrespective of race. Such admiral and radical thinking from a much admired mathematician of his day was built on and Condorcet further argued that as nature was perfect so was the man who lived in harmony with nature's laws.

The argument then pivoted on a fundamental disagreement as to whether this meant that primitive man, who lived closer to nature, was more perfect than modern man, or whether through civilisation man was getting closer to perfection thereby passing the key to such utopian dreams to future generations. The latter was Condorcet's view and Ovid's 'Golden Age' had become Condorcet's 'Golden Future'.

Malthus, perhaps spurred on by his faith, for he was a priest in the Church of England, responded to The Marquis de Condorcet's proposition around the 'perfectibility of society' through civilisation, rationalism and education, in his essay on *The Principle of Population*, which he was to review and hone six times over a twenty year period. This piece of writing was to fall upon Condorcet's deaf ears for he died aged 50, four years before it was published. Nevertheless this treatise gained a wide audience and proved extremely controversial causing many, including William Godwin, who was more aligned to the Condorcet's school of thought, to take up the argument.

As with all such matters the participant's upbringing and education would have heavily influenced the discussion and even what the discussion was about. Condorcet offered his brilliant mathematical and logical mind applying it to his economic theories and arguing that there were no limits to man's improvement and provided matters of race, gender, religion and culture were put to one side then man's destiny and achievement of a utopian society was in his own hands; perhaps the birth moment for the modern secular age. Malthus responded, his thinking probably framed by his belief in the fall of man, by expounding his theories based on his studies into behavioural economics.

In his essay Malthus made the following observation on the state of man and concluded;

'Must it not then be acknowledged by an attentive examiner of the histories of mankind, that in every age and in every State in which man has existed, or does *now exist that the increase of population is necessarily limited by the means of subsistence; that population does invariably increase when the means of subsistence increase, and, that the superior power of population it repressed, and the actual population kept equal to the means of subsistence, by misery and vice.'*

By misery and vice he meant war, famine and disease. Interestingly it was Charles Darwin nearly half a century later, who extended these conclusions into the field of biology and his theory of natural selection, although diplomatically no reference was made to man. Malthus was undoubtedly very pessimistic about the future of humankind, and history has proved him wrong at least in terms of the timescale of events he outlined. He had very clear and practical ideas as to how to limit population growth. These included delaying marriage until the union was financially viable and demanding celibacy in the years before marriage. His critics argued that the morality of this solution together with his support of the gradual withdrawal of the Poor Laws made him anti-poor; a position that Malthus vehemently denied.

Subsequently in Western society his approach, albeit based on a rather different rationale, is precisely what has happened and people are getting married and starting families later in life and practicing birth control with the consequence that population is in a downward trajectory. Even for less economically developed countries there is a well grounded argument in circulation that a month of sexual abstinence can have a hugely positive effect on levels of HIV infection and of course birth rate; Ramadan is cited as the experimental evidence.

The question for us is whether man will slavishly follow the classic population curve preordained by biology until such time as 'subsistence, misery and vice' curtail and collapse it, or whether there is a more 'utopian' way whereby the application of education and progress can stem and make it sustainable? The Malthus versus Condorcet argument remains unresolved and urgently relevant.

The United Nations in its *State of World Population 2009* paper, issued with the then upcoming Copenhagen Climate Change Conference in its sights, made some interesting observations. It noted as its reason why population level issues had not been addressed by international and governmental bodies that;

'...population growth and what, if anything should be done about it, have long been difficult, controversial and divisive topics.'

The key of course to population growth lies, fairly or unfairly, with women for they bear the new beings that replace or add to the population. By number women represent the poorest on the planet, the least well educated and most susceptible of the genders. Equally it is clear that in countries and regions where the standard of living is better, where women receive education, contraception is widely used and women conceive later in life then population growth has stabilised or is even negative. Where the converse conditions occur, as in sub-Saharan Africa and Southern Asia, then population growth is fast and furious.

It is therefore not surprising that the United Nations along with many NGOs and charities has long recognised that any solutions to population and poverty, rests primarily with women. Consequently their focus has been on assisting women in the management of fertility and the provision of food. This gives great cause for hope for whom better to cope with crisis than the most hardworking, compassionate and multi-tasking gender of our species? What does come through as a clear conclusion to the UN's considerations is that having skirted around dealing with population numbers as an issue in its own right, it concludes that climate change is exacerbated by population growth and population stabilisation will help reduce greenhouse gas emissions albeit like everything to do with population this will take time.

Two hundred years have passed since the Condorcet/Malthus debate and still people continue to take the same sides with some believing in the ability of man to solve this issue and deliver a sustainable population through the civilising influences of education and science, whilst others are reconciled to the inevitability of population collapse. Perhaps the real truth is that Malthus was right, albeit rather later than he envisaged, and we are inextricably heading for a population crisis and crash that will be at least in part cushioned and mitigated by Condorcet's trust in man's ability, using the tools of the enlightenment, to soften the eventual landing.

Of course this is not a neat matter and some regions of the globe will experience the best of Condorcet's optimism, whilst others will and indeed are already suffering the full horror of Malthus's pessimism. For us it has to be this mitigation argument that provides the trumpet call to all those directly engaged in, and who work tirelessly to bring educational and health care to all lands.

The three solutions

There is clearly no single solution to this intractable problem, but there is a whole armoury of means available to fight and secure a bridgehead to mitigate the global and local problems of population explosion.

In the 1970s the issue was verbally linked to a class system and the West developed its own nomenclature; First World, Developing World, Third World and Fourth World. The Fourth World was considered to be those parts of the world where even 'lifeboats' of aid from the First World could make little impact as such was the size and level of the problem of poverty that nothing could be done to alleviate and change their lot. That class nomenclature was soon dropped and we transmuted into using definitions such as Developed World, Developing World, but they too now seem inadequate as the so called Developing World of China, India, South Africa and Brazil move into controlling mode of the world's economy. Just as we have accepted that equality of race, gender, and religion are essential for an ordered world so that equality should extend to nations. Each has its own particular contribution to make to the global economy whether that be in the form of people, invention, minerals or food. Once we can free ourselves from pigeon holing each other then each nation can work together as equals and focus on education, birth control, equality of gender and greater security for history holds the answer and provides the lamp for our feet.

What history tells us, very clearly and in unison with our biology, is that where life becomes less risky then the birth rate falls away; and risk reducing factors are education that enlightens, birth control that gives women greater control on fertility, and security that dampens the fear of the future.

We need to swing the argument in Condorcet's favour whilst recognising the truth of Malthus that to do nothing, or to do too little, will cause a species to enjoin huge suffering and pain; the problem for us is that species is us.

Population is about people

This aspect of human civilisation is beyond government control unless it is particularly authoritarian and even then, as with the China experiment, it leaves behind it a whole raft of social issues; tampering with population en masse seems to be a bit of a Pandora' box. Incentives to voluntarily limit family size have made little inroad on the figures and however the matter is approached it always comes down to the individual level. This was recognised by the 1994 International Conference on Population and Development hosted by the UN when it was agreed,

'that population is not about numbers, but about people.'

Whilst there are perhaps no ethically acceptable ways of enforcing population control we have already referred to two devices that can have a major impact on its restriction. These are education and the tools of birth control. In many African countries the availability of contraception devices is limited to less than 10 percent of the population. Inevitably there are a large number of unplanned and unwanted pregnancies all adding to the impoverishment of the people and the building of pressure on food supply. Simply by making such devices available to the whole reproductively active population would markedly reduce population growth to the benefit of the individual and all. Even the central machine of the Roman Catholic Church is slowly moving to that conclusion and leaving its unsupported and, some would argue, questionable and even damaging theology on this matter behind.

The great conundrum of human behaviour as described and asked by many is whether increasing food production encourages population growth or alternatively whether population growth encourages greater food production. Answered simply the response to the first question must lie with the biologist who would argue that the population will grow and flourish in response to plenteous food supply, whilst the answer to the second question will rest with the human species' humanity that drives him to provide for his fellow kind whether it be for profit and power or for purely philanthropic reasons.

Outside of the occurrence of some single catastrophic event what we do know is that by 2050 the world population will have reached 9 plus billion. Whatever we do now will have little bearing on this inevitability, but what we do now in terms of providing education, means of proper existence and the tools of birth control will have a profound effect on the numbers to follow and make this moment in our species' existence a bulge on the graph rather than the beginning of a catastrophic and quick end.

But we need to listen to a word of warning for reducing population can be as hazardous as its increase. As the world moves to a declining population so the needle marking average age will rise and for a period the globe will have to cope with a disproportionate number of old, infirmed and non-productive members all requiring support from the dwindling numbers of economical active people, but that is for the future; we have enough on our plate right now.

The question then remains as to whether we can support the nine billion plus population in the short-term and then what level of population do we need to work towards that can be sustained over the long-term? That is an agricultural matter now being made seriously complex and uncertain by climate change.

HOTTING UP

The die is cast

It is not possible to talk about any matter that lies ahead without including climate change in that discussion for it will have a profound effect on everything we do and what we can achieve. No person or facet of life will escape its inevitability as the die is already cast.

As a single young man living near Canterbury I would sometimes attend the cathedral Evensong at 6.30pm on a Sunday evening often having worked through the day. Not many did and we would gather in the chancel attended by members of the Chapter and the Cathedral Choir. Such buildings own a unique atmosphere in the evening quiet and their emptiness and it is easy to sit in history and join those, who in prior times had sat in the choir and participated in or watched on, as the familiar service had been recited over many succeeding centuries. On a particularly dark, cold and very wet Sunday evening in January 1974 I entered the ill-lit Cathedral Yard to see a couple trying to gain entry into the building via the south door. I walked over and led them, against the now horizontal rain, to a small side door at the crossing which was used for access for this service. As they entered the Cathedral and removed hat and rain hood, they turned to thank me and I recognised them as Dr Coggan and his wife. Two weeks later he entered by the West Door and was enthroned in greater splendour as Archbishop of Canterbury and Primate of All England. The Dean, Ian White-Thompson, preached that evening, and whilst I do not remember much of what he said, I do recall his saying that *'loving a child is not enough, for you have to demonstrate that love for it to have meaning'*.

The fact that the climate is changing and temperatures are rising and that this is a threat to mankind and many other species is not seriously disputed, but knowing that and being concerned about it is not enough for we have to demonstrate our concern by doing something about it. There is a whole plethora of articles and books analysing the situation in terms of cause and effect and yet when it comes to actually tackling the issue the pages are few and often the listed ideas unconvincing. Nevertheless there are movers and shakers engaging with the issues surrounding climate change and we must work with them for if we ignore this matter we could well be doing so at our own peril or our children's peril.

No respecter of persons

Whilst there is universal acceptance of global atmospheric temperature increase, what is debated is whether this is caused by man's activities or whether it is simply a natural occurrence as has been evidenced before in this planet's long history. The majority of scientific work points to the former whilst a vociferous minority correctly claims that science is not always infallible and then goes on to argue that other evidence is more compelling. Of course it could well be that both are right and we have a coming together of man-made and natural phenomena, which are at least for the time being accelerating climate change and world temperature increase.

Undoubtedly the science is immensely complicated and discerning the cause is difficult enough, but when it comes to discoursing on the future outcome we walk into a fiendishly complicated web that, in truth, is probably beyond our understanding and is likely to largely remain so. However, from a pragmatists view point, whether scientist or non-scientist, it must be good sense to assume the worst and harvest the benefits of our actions should such a prognosis be found to have been an over exaggeration. The risk of doing too little only to find the less favourable predictions of some science comes true is tantamount to madness.

In summary form and distilled down climate scientists are predicting, with more than 90 percent confidence, that climate change will lead to higher maximum temperatures with more hot days and heat waves over nearly all land areas, higher minimum temperatures with fewer cold days, frost days and cold waves, and intense precipitation events that will increase floods, landslides, avalanches and mudslide damage with more soil erosion and flood run-off.

How climate change will affect our biological existence as a species is complex. In essence large parts of the landmass will become less habitable as shortage of water and desertification reduce its crop and animal carrying capacity and urbanised units will become less viable and able to support their populations. In addition the effects on health could be near catastrophic particularly for the very young and old; look what happened in France in the summer of 2004 or the Moscow area in 2010 as a taster.

This will engender the need for human migration to more favourable areas including some marginal temperate regions made the more habitable by the temperature rise. Migrating refugees, as always, will journey with suffering and their paths will take them to points of potential conflict along its way. The world will in effect be increasingly split over a period of decades with a massive flight to the north and a lesser southern movement of peoples with the full gamut of political, social and welfare issues this, for certain, will generate.

The economic fallout from continuous global warming has been assessed by Gordon Brown as resulting in an economic crisis worse than the Great Depression and the two World Wars combined. That is a huge statement from a naturally understating Scotsman, who further recognised that on top of this there would be hundreds of thousands deaths every year due to floods and droughts. In effect his warning was that unabated climate change is likely to cause global GDP to fall by 20 percent. This led to his UK government making a commitment to set a target of reducing UK emissions by 34 percent by the end of 2020 and 80 percent by 2050 as against 1990 levels. The 2010 incoming Conservative administration in coalition with the LibDems has continued to support these goals with promised investment in green renewable and nuclear energy.

The biological and economic fallout are inextricably linked and as always the first to suffer will be the poor and as is the case so often it is women who will suffer the most. It is estimate that 70 percent plus of the poorest of the poor are women and that currently 75 percent of all environmental refugees are women. But the consequences will not stop with today's poor and if matters are just left to follow nature's red in tooth and claw agenda then we will suddenly realise that 'we are all poor now' for climate is no respecter of persons

Sharing the blame

The business, and perhaps most logical, approach to climate change is to create a balance sheet. On one side of the Climate Change Balance Sheet there would be the list of sources and respective quantities of man-made or man-influenced carbon emissions being discharged into the atmosphere each year net of any carbon capture. Assuming we need to reduce these by adopting the UN figure of 80 percent by 2050 in order to bring a halt to increasing the parts per million of carbon dioxide in the atmosphere, then we have a real target figure for carbon emissions; an 'allowable figure'. The other side of the balance sheet would therefore consist of the 'allowable figure' for each individual source of emission with the total savings delivering the balance. Whilst simplistically each source would be cut by 80%, there may be priorities such as food production which require preferential treatment.

The split of anthropogenic, or man made, carbon dioxide generation is shared out between the main sectors as follows:

Energy generation	32.6%
Industry	16.8%
Transport	14.0%
Agriculture	12.5%
Property: Residential/Commercial	10.3%
Land use and biomass	10.0%

Waste Disposal   3.4%

To this list must be added deforestation which adds to carbon dioxide emissions by reducing their absorption. This activity alone is estimated to account for 20 percent of man-influenced carbon dioxide emissions.

This data enables us to quantify and measure the solution and in so doing to have a framework for debate; hopefully for not too long for once the broad targets have been identified it becomes increasingly urgent to finds means of delivering them. For the purpose of this particular journey in thought just four of these emitters are considered representing 80 percent of the emissions; they are energy generation, deforestation, industry and agriculture.

From black to green

The British Isles' energy history has been a pretty good forerunner of the progress in this field that many other countries have and are coming to follow. In the years of empire and through to the years following the two World Wars coal was literally the country's bedrock. Coal was the source of just about all our energy and during that age it was being mined in abundance from the great coal fields that scarred Britain in a great swathe from Scotland through the North East and Yorkshire down through Nottingham and the Midlands and into South Wales. At the end of that century Great Britain and the USA far exceeded any other nation in the production of this black fuel that drove the country's manufacturing base as it busied itself adding value to colonial and home sourced commodities and raw materials. It was coal fired power stations that generated electricity that laced its way from pole to pole to even the remotest parts of the land.

Coal became intensely political as the age old battle between miners, seeking a living wage, against the coal owners, bent on creating wealth, was transferred to the state with the nationalisation of the industry in 1942.

Whilst farming in East Kent I watched on as the pits in the area were closed until there were just three left. Many of our harvesting and packing staff were miners' wives and we looked out over the Betteshanger Colliery as it rumbled and hissed around the clock. The Kent miners were literally a breed apart for during the years of depression their fathers and grandfathers had walked from South Wales, Yorkshire, Durham and Scotland to find work in these new pits. Hardship was a part of their gene pool and whilst by nature they were generous to a fault they could be harder than nails when irked.

The politics of the Thatcher government included a deep-seated resentment against the miners, who had used, or some would say misused, their industrial muscle in the earlier 1972 and 1974 strikes when Edward Heath was Prime Minister. This long standing vendetta was mixed in with a review of the coal industry that led to the March 1984 strike, which was to last for almost a full year.

The Kent miners stood shoulder to shoulder and throughout the year never wavered in their determination to fight their cause to the bitter end even when other mining areas began to buckle under the hardships that such a long strike inevitably brought. For the Kent miner it was largely an uneventful strike, relatively peaceful and good natured. Their focus was on survival and the many miners' wives we employed worked hard to maximise their earnings whilst remaining in complete support of their men folk. We provided free vegetables whenever possible, which were collected and distributed by the union and despite being in the middle of a colliery area and well known to that community we never suffered any measurable loss of crop from our easily accessible fields. Despite this determined, but peaceful existence, there was an insidious change in events when without warning lines of police coaches could be found parked up in country lanes or driving through the area. I would count thirty to thirty six such minibuses each containing twelve or so uniformed policemen. Not only the miners, but those of us living and working in the area felt threatened and resented their presence. There was no reason for such a police presence when a handful of bobbies at the pit gate would have done the job; it was the state practicing raw intimidation in our part of this green and pleasant land. The rest is history and the strike failed and the mines closed and one Sunday morning I watched from the neighbouring field as unannounced the pit frame was felled by carefully placed explosive charges.

It is easy to forget how recently alternative fuels to coal were introduced when we are already talking about their demise. Oil, like the coal it was to partly supersede, also enjoyed its own history made personal by characters from the first half of the last century. The importance of this fuel only became evident to politicians and business during the First World War and the prime source of this new fuel, which flowed instead of being hewn, was a newly formed country. This fledgling nation was ruled over by the House of Saud, who gained power in 1902. They were a family happier riding to camel in the desert rather than holding court. With oil beginning to flow from its shores the Americans and British suddenly took a new interest in this dry and patched piece of desert. How they did this was to have major repercussions on their future relationships and prosperity.

President Franklin D Roosevelt, aboard the U.S.S. Quincy, invited the Saudi King to meet with him. The King was transported to the rendezvous by a US warship on which he and his entourage, including one hundred accompanying sheep, promptly camped down on the deck. At the meeting of the heads of these two states all evidence of alcohol and tobacco was removed from the Ward Room in deference to Islam and the visiting King. The meeting was cordial and accompanied by Arabic coffee brewed by the King's servants. When the King, who was getting old and rather stiff, showed an interest in Roosevelt's wheelchair he was promptly presented with the spare one stowed on board. These two very different men from very different cultures immediately formed a bond and although Roosevelt was to die two months later the relationship of nations had been forged.

When three days after this meeting Winston Churchill met with the King things were very different. The Grand Hotel du Lac, south of Cairo, with its rather indulgent and louche ambience set completely the wrong tone for the King, a situation only exacerbated by Churchill's insistence not to have anything to do with any kind of abstention in deference to the King's principles; wine and spirits accompanied by his famous cigars flowed freely. Of course the King was much offended; a situation that was made only worse when later he received a gift of a Rolls Royce. This car epitomised the pinnacle of British engineering and luxury. The problem was that it was a right hand drive model suited to English roads, so that the chauffeur had no choice but to sit at the right hand of the King; a position that was nothing less than an insult to such an exulted personage. The car was left garaged and unused.

The net result was that the House of Saud adopted the USA as their partner and mentor as they joined the 20th Century and cast off their more mediaeval existence and began to exploit their natural wealth of oil and to become the influential and prosperous nation of Saudi Arabia that we know today.

However, in terms of electricity generation it was natural gas that really led the charge to fill a large part of coal's shoes as the prime means of electricity generation. It was readily available off-shore in the North Sea and had the benefit of producing just half the level of greenhouse gases that black coal produced and 30 percent less than oil. However, not everything was positive as natural gas is methane and methane is eighty times more potent as a greenhouse gas than carbon dioxide. As a consequence over 14 percent of anthropogenic greenhouse gases are derived from methane or natural gas.

In parallel was the equally controversial development of nuclear energy with its zero greenhouse gas emissions, but it brought with it serious concerns regarding its financial viability and the back-of-the-mind nagging fears as to its safety, particularly after the 1980's scares at Three Mile Island and Chernobyl. Fears that have been further underscored by the 2011 Fukushima disaster in Japan.

Finally the birth of green renewable energy is largely of the 21st Century and whilst still a very small contributor its course is set to fuel the globe in the second half of this century as oil in particular phases out. The sun has generated the fossil fuels of coal, oil and gas that we have so greedily devoured and each day it continues to dispense vast quantities of energy upon the earth's surface. In effect we have, unharnessed, a limitless amount of energy available to us wherever we might be on the planet. The sun provides direct energy in its light and heat, creates wind, generates rain that feeds the rivers and interacts with chlorophyll in plant life to produce biomass. Even the moon plays a part in the gravitational generation of tides. Indeed all our energy comes from outer space.

In Europe, and possibly the world, Germany has taken the lead in terms of the revolution in energy generation by declaring its hand and after the Fukushima disaster announced its intention to close down its nuclear plants and replace the electricity they produce with green energy generation.

We only need two power stations

The challenge facing scientists and generators working together is how best to capture this free flow of energy and transform it into usable electricity. This, combined with carbon capture techniques resulting from the burning of fossil fuels and scheduled to come on stream by the mid 2020s, raises real hopes that power generation can become significantly cleaner if not completely carbon free. We only need a couple of power stations; the sun and the moon.

Blunt axes

Deforestation of this planet is not a new phenomenon. What is new is the scale and rate of that deforestation. Go back to the glorious days of Merrie England and its celebrated times when it destroyed the Spanish Armada and plied the seas in search of new lands far and wide. Those battles were won and colonies gained at a cost, and that cost was the great oak greenwoods of England and the decimation of the tree population. The flip side to this was bought home to me when I was visiting the town of Tortosa on the Spanish coast just south of Barcelona. The town has been distanced from the sea by the fertile Ebro Delta, which now supports a great deal of prosperous agricultural activity. As I walked the fields my host reminded me of our mutual history as he explained that the delta had been formed in a matter of a few hundred years following the deforestation of the hills behind Tortosa during the days of building the great ships of Spain that plied the oceans, discovered the South Americas and went to conquer England as the Armada. Today, mechanisation has upped the rate of such brutal felling round the globe and the rewards are great; for this generation, as with oil and coal, is harvesting the past faster than it lays down the future.

From our table it is clear that if we could move to sustainable forestry whereby wood felled equated with annual tree growth then immediately we would reduce our carbon emissions by 20 percent. Not only that, but ecosystems would be maintained and biodiversity secured, for much of life's variety is corralled by the belt of tropical rain forests that still stretch round the planet. Fresh water supply would be enhanced and this carbon sink would gurgle happily and helpfully away. Of all the options to reduce carbon emissions this one act seems to have so many beneficial outcomes that to many it is a no-brainer.

Of course with population growth there are real pressures on timber as a fuel for cooking and warmth and for construction, and no human has the right to deny another such basic needs. It follows that reverting back to having sustainable world forests will require a mix of felling and planting policies and activities that best match the needs of those for whom the tree-lands provide a home. On the other hand there is huge pressure on the land upon which these forests grow and when 50 acres of South American forest are laid waste to support a single head of cattle to feed the burger outlet in New York or London then there is every reason to call a stop. Whilst the complexity of all these issues is immense the imperative is simple; we must blunt our axes.

Industrial migration

There has been a major shift in energy consumptive manufacturing from West to East with China in particular picking up the gauntlet of supplying the world with consumables often deploying energy intensive manufacturing methods as with steel and fertiliser production. This has resulted in a corresponding carbon shift exacerbated to a degree by the fact that overall Chinese manufacture is less carbon efficient than the comparable processes it has replaced in the West.

In the early 1990's I was Group Managing Director of several companies all linked to India with its heritage in the East India Company. One of the company's advisers was a quietly spoken Parsi who had been Chairman of The Times of India, a director of Skoda, and who then lived in Singapore and acted as one of the President's advisors. He once observed that whilst Russia had gone for political change first and foremost with the fall of communism and would then go on to develop its industry and trade once the dust had settled, the Chinese would go for the development of trade first and only then begin to address political change. With a slight smile and glinting eye he concluded 'that is why China not Russia will drive the world economy.' He has been proved right for historical China was the birth place of entrepreneurism and trade.

As a consequence China is now the world's largest carbon emitter producing 6.5 billion metric tonnes per annum with USA, recently overtaken, on 5.8 billion tonnes. The arguments around justice and equity come in to play when those emissions are converted in per capita sums with China on 4.8 metric tonnes per person and the USA on four times as much at 18.7 tonnes per person. The UK sits at 8.6 tonnes of carbon produced per person each year. This figure has diminished since 1980 when we were responsible for nearly 11 tonnes per capita, however much of our energy intensive industry has now gone abroad to be replaced by imports, which inevitably distorts this apparent story of success and nullifies most if not all of this apparent improvement.

Agriculture

Apart from the emissions we see from exhaust pipes or factory stacks and cooling tower as they belch endless plumes of smoke and steam there are also some less obvious and less familiar sources. Agriculture harbours one such time bomb. Bovines are ruminants geared to digest cellulose, which is ideal for these grass eaters. During food digestion, aided by many microorganisms living in their rumen, they produce methane gas, which they then breathe out into the atmosphere. The generation of this extreme greenhouse gas is of particular concern and The United Nations Food and Agriculture Organisation have cited this matter as being of major import to climate change as the global consumption of meat and animal products increases. The Food and Agriculture Organisation has even suggested the possibility of levying taxes, applying fees or cutting government subsidies on cattle, concluding;

'continued growth in livestock production will otherwise exert enormous pressures on ecosystems, biodiversity, land and forest resources and water quality, and will contribute to global warming'

Of course agriculture not only produces natural greenhouse gases, but also emits large quantities of carbon dioxide as it consumes oil to till the land, to treat the soil and crops with chemicals and fertilisers, and then dries and processes the harvested grains or produce. Farmers are deeply conservative and a cautious breed when it comes to comparing tried and tested husbandry method with new advances. With an active and sometimes extreme variable such as the weather always knocking at their door, it is understandable that to add any further uncertainty to the business of growing is eyed with a degree of caution. This is made the more so as most farmers live a solitary life and most business decisions remain unshared and become completely identified with the person; success or failure in business becomes success and failure in life. But farmers will respond to new situations and can change.

Such a stimulus for change will be the upward movement in oil prices that undoubtedly lie ahead. In 1960 the price of petrol at the UK pumps was 5.18p per litre. Since then it has gone up in each decade to 7.15p, then 26.39p and on to 40.92p in 1990 before doubling to 80.84p in 2000 and reaching 140p in 2011. If we extrapolate this trend line then by 2020 we would be looking at circa 180p per litre, which may seem to be light in view of progressive taxation on carbon and the possibility of peak oil exacerbated by increased world demand. Link these pump prices, which form part of the understandable everyday currency of our lives, to the barrel price of oil over the same period and a slightly different picture emerges. Here the inflation adjusted barrel price of oil has remained remarkably constant at between $20 and $30 dollars right up until 2000, despite blips caused by the Yom Kippur and Iraq/Iran wars during the decade following 1973. However, in the new millennium prices consistently and quickly increased to peak at nearly $150 per barrel in 2008 before falling back as the world economic recession set in. The cause of this inexorable rise was principally the growth in Asian industry as it began to claim its role as manufacturer to the world and put pressure on oil supply.

Using industry estimates peak oil is forecast to occur some time within the next thirty years depending on different discovery and consumption rate scenarios. The likelihood of the $200 per barrel price this decade is therefore a real possibility, some analysts say a certainty, as the world recovers its economic poise and once again drives for growth. The importance of the $200 per barrel price is considerable, for it is a real and physiological point when many of the dynamics of industry and agriculture change.

In the oil based world of intensive Western farming it is forecast that energy costs of this magnitude will make lower input organic farming more profitable than conventional, or oil based cultivation, of crops; and that is without any organic premium price advantage. In other words the balance tips and the cost of achieving extra yield from high input farming exceeds the returns that can be achieved. Of course new technologies will impact on this equation to an unknown extent, but the underlying message is clear; the next generation of farmers in the West may well find the profitability of sustainable agriculture more attractive than their counterparts of the current generation.

It is sometimes difficult to envisage such a change in attitude in such a conservative fraternity. But farmers can respond, and respond quickly, to change if there is good reason to do so. In the late 1970s there was a brief spell when the prevailing wisdom was to burn the straw in the field where it lay after the cereal crops had been combine-harvested. It was argued that these huge, and sometimes frighteningly exciting infernos, cleaned the fields and allowed for easy cultivation ready for autumn sowing of the next crop. In fact the key reason was a fall in demand for straw as stock farms moved to less labour intensive animal bedding systems. Field after field was burnt along with many hedgerows leaving mile upon mile of blackened land. Tractor drivers returned home looking like coal miners coming off shift and the farm took on an air of sterility. Within a few years the practice ended as quickly as it had begun much to the relief of the fire service, as the penny dropped that such method was anti-social, harmful to wildlife and left the soil short on recycled organic matter. Overnight, instead of setting the world alight, choppers were fitted to the rear of the combine, which then returned the straw in short pieces to the soil. These cut pieces quickly decomposed helping to build soil structure and fertility. The farmer, the lady with her washing hung out to dry, the tractor driver and the insects and beasts of the field all regained a degree of harmony and a modicum of sanity once again prevailed.

Similar examples can be found with the advent of chemical sprays to protect crops from weed competition, fungal diseases and insect pests. In the 1960s the armoury of such chemicals was small, but their use became routine. DDT and Aldrin were sprayed alongside slightly less noxious chemicals often with little protection for the operative. Chemical company representatives continually turned up with new and exciting listings of chemicals to incorporate in the soil, to apply as granules around plants or to spray on bare ground or over the standing crop. Some crops, like Brussels Sprouts, which seemed susceptible to just about everything and stood in the ground for many months vulnerable to attack, would be sprayed or treated anything between fifteen and twenty times in order to deliver the visually perfect product that the supermarket demanded. The days when my grandfather would reputedly place cooked caterpillars from his home grown cauliflower on the side of his plate were long gone.

One farmer I knew, who was at the vanguard of this new chemical movement, happily maintained that spraying his wheat with chemical was potentially no more hazardous than spraying it with coffee. Such was the farming parlance of the day and decisions to spray a crop were made without recourse to any scientifically developed parameters, but rather on the whim and judgement of the farmer, often largely unschooled in such matters and after having wandered into the field to examine a small sample of plants. Perhaps hastened by the widely publicised fall out from dipping sheep in organophosphorus and the terrible consequences on those exposed to this chemical, the industry began to take agrochemicals more seriously and today it is under much improved management with tighter controls and their application largely rests with professional contractors. These examples demonstrate that even one of the most conservative of communities can change their practices and when necessary change them quickly.

As with the forests and industry it would seem that agriculture is going to require a commercial excuse or reason to make the change to a low carbon existence. All the well intentioned sentiment in the world about such matters will not overcome the relative short-termism associated with dealing successfully with the annual cycle of most farm crops. 'If it's not broke why fix it' will always gain greater applause than any appeal for less tangible goals that may anyway appear to be several seasons away.

Navigators of the future

The arguments around climate change are important, for to drive through these seismic changes to carbon usage is beyond government alone and requires acceptance and the harnessing of the will of the people. Sadly the white knight of science that rode on to the scene in the 20th Century is now looking somewhat battered and tarnished. Nevertheless it is a key player despite the damaging Climategate debacle at the University of East Anglia and the problems with data from the Intergovernmental Panel on Climate Change 2007 Report.

But such occurrences should not be used to besmirch the scientific contribution and there has never been a greater need for scientists and those who interpret their findings to speak with clarity, transparency and objectivity. Governments, industry, commerce and the people must be able to rely on and trust these findings if meaningful progress is to be made; they are no less than the navigators of our future.

Sadly the reality is that as a result of these failures in scientific research together with the collapse of the Copenhagen Conference in December 2009 we now have a deeply sceptical public, who are unsure what to believe and who may be grateful to grasp any straw that might unpick the unpleasant truth of what lies ahead, however futile a gesture that might be. Indeed as the world's economies double dip and governments and people become preoccupied with getting through days and weeks rather than months and years so climate change has fallen out the media and no longer figures to the extent it did even twelve months ago.

But the uncomfortable truth is that that climate change is happening and it and its consequences cannot just be wished away or ignored. I was somewhat taken aback by a researcher working in this field who in defence of the slow rate of progress they were making in their research quoted their Head of Department as saying that the 'owl of Minerva flies at dusk'. I did not fully understand the saying so did the necessary research and discovered that it meant that the owl of Minerva, representing wisdom, flies at the end of the day; or put more simply wisdom only comes from hindsight. If we are to believe this then everyone is a don and full of wisdom about something; the problem is that it will be too late to do anything about our predicament, wisdom or no wisdom. If that is the culture within some of our universities it does not augur well and seems a long way from the great early days of science when the likes of Newton, Halley, Hooke and more beavered away with rather greater determination.

Recently I revisited Woolthorpe Manor just off the A1 in Lincolnshire and for an hour relived Isaac Newton's life. Outside in a small orchard was reputedly the famous apple tree and then inside, the room where he was born, and on to another room where he devised and thought through much of his early work. Newton's First Law and his Second Law and then his Third Law of physics, together with his theory of gravitational pull, all of which still hold good to this day, were but a small part of his life's work, which also encompassed mathematics and optics. Then there was the information note informing the reader that in fact he spent the greater part of his time on religious matters, writing a considerable number of pamphlets and works expounding his theological views. These views, which he held so strongly and worked so hard to set out, are long forgotten, but his science, which was inspired and life changing, lives on. It is exactly this sort of insight with its simplicity of genius that is required in the field of climate change today.

Of course 'Newtons' don't come along very often, but with a much larger pool of researchers in the global arena there is every chance that one already exists and we must seek them out and listen for their voice. As a man of the modern age I was struck by the remarkable fact that Newton's endeavour came out of an isolated farmhouse from a man whose father was illiterate and who had died before his birth. As I was about to leave I read Newton's own words written towards the end of his 84 years;

'I do not know what I may appear to the world, but to myself I seem to have been only like a boy playing on the sea-shore, and diverting myself in now and then finding a smoother pebble or a prettier shell than ordinary, whilst the great ocean of truth lay all undiscovered before me'.

For Newton there was no waiting for dusk and owls to fly; life, even in those far off days, was far more urgent and exciting.

Market or man

Science and technology alone cannot resolve the climate change issue for if, as generally agreed, it is caused by people, then people have to change as well. But people will only change if there is a degree of underlying equality and a feeling of fairness or justice. Many thinkers have raised the big climate change question; should those countries who have enjoyed the benefits of an early and polluting period of industrialisation now shoulder a large part of the costs and consequences of those countries now aspiring to the West's standard of living by going through there own polluting moment of industrialisation.

Fairness is a human concept that is alien to biology, and climate change and its consequences for these are biological issues. That is not to discount the human search for justice, but justice is not a game of cricket operating under inviolable rules and in a gentlemanly spirit. Justice means exercising concern and deploying its outcome for the benefit of our fellows in order to mitigate the harshness of natural laws. Justice is a hard taskmaster for the bestower and the receiver and requires considerable wisdom as it digs deeper than the more superficial craving for fairness. Therefore the answers to these questions are not easy and the options are often portrayed as either going along with authoritarian government imposing change on the people versus allowing individuals and the markets they create to dictate the solution. Perhaps E F Schumacher best answered this dilemma back in 1973 in his seminal work *Small is Beautiful* when he wrote

'In the affairs of men, there always appears to be a need for at least two things, simultaneously, which, on the face of it, seem to be incompatible and to exclude one another. We always need both freedom and order.'

Government – plenty of tools but no hammer

It is generally assumed that governments hold the key to delivering substantial reductions in greenhouse gas emissions. That is the easy solution and undoubtedly when a government brings in legislation it delivers immediate and major change, but the imposition of new laws requires a large degree of support on the ground in democratic countries and even under more authoritarian regimes such litigation has to meet the demands of particular interested and influential parties. Neither route is easy, especially when living standards or powerful personal financial interests are adversely affected.

A few years back I was interested to read an unpublished paper by an agronomist who had spent much of his working life in East and West Africa. It was in the early days of the climate change debate and I was initially taken back by his conclusion, simply because I had not thought of it in these terms before and had not expected a financial solution from a field scientist. He advocated that taxation was the prime lever and means of reducing emissions.

The debate has moved on and the role of taxation is now widely recognised as a part of the armoury required to suppress energy consumption; although if we take the example of the UK, it has had limited success simply because the people have not accepted the argument. If so called 'green taxes' could be solely and convincingly applied to investment in green projects then they might attract wider support, but such is the scepticism towards government, that even that might not significantly widen their narrow base of support. As a consequence, the inability or unwillingness of governments to impose tax on aviation fuel and thereby reduce its spiralling and polluting usage remains a lasting epitaph to the failure of such governmental endeavour.

Where governments have been more influential is in supporting fledgling renewable energy technologies and projects financially. Any new technology goes through a period where innovation costs and the financial penalties associated with small scale production and high marketing costs weigh heavily against its survival. The world economic recession at the start of the 21st Century has temporarily slowed the pace of such initiatives, but nevertheless renewable energy schemes continue to come on stream with Europe's Desertec initiative targeting 15 per cent of Europe's power to be generated in North Africa using solar farms, and the UK wind farm generation project targeted on producing 25 percent of the country's needs within a ten year period. The problem is that at present energy from coal costs about 6 cents per kilowatt hour to generate whilst wind power generated energy is about 12 percent more expensive and solar power is approximately 300 percent more costly. However, as manufacture is scaled up and technically advanced so the costs of production are now responding and coming sharply down. This will undoubtedly further stimulate the market and bring in once marginal schemes.

In 2010 alone 10 gigawatts of solar cell capacity was manufactured within the global market, representing a 50 per cent increase on the prior year. With a financially constrained Europe the initial government subsidy in support of these industries is falling away and as a consequence production is moving east to low cost Asian manufacturers and west to the energy hungry USA market which has retained its subsidies. This just accentuates how fluid even manufacturing has become in this global village and how influential government subsidies can be in short term market distortion even if the waters of the market eventually close over again.

A further and important example of direct governmental impact on climate change is their efforts to slow the rate of deforestation, particularly in the great rain-forest regions of Africa, Indonesia and Brazil. The great Rain Forests of Brazil are the most frequently quoted example where each year an area of around 10,000 square miles has been lost as a result of tree felling; Wales is circa 8,000 square miles in area. Over the past few years this has been radically reduced by government action plus a handful of hectoring environmentalists, who not infrequently have paid with their lives. In 2010 the area lost to these forests was 2,500 square miles; this still leaves much to do but the trend is in the right direction and we all stand in debt to those eco-martyrs.

Of course with the global financial crisis, which we will undoubtedly continue to experience through this decade as economic power migrates from West to East, the need for private funding will become more urgent if we are to maintain momentum on protecting these natural carbon sinks. Many argue that stopping deforestation is the single most important thing we can do in our efforts to combat greenhouse gas production and climate change. Currently 20 percent of the world's carbon dioxide emissions are attributable to the loss of forest, which would otherwise have absorbed these gases.

So far governments have earmarked $4 billion dollars to safeguard the world's forests and there is pressure to use the carbon trading schemes to help fund this endeavour further. In 2010 the carbon trading market was valued at $142 billion despite immediate recessionary pressures and many expect it to become the single largest commodity market in the world. These traded credits are issued by the United Nations to greenhouse gas reducing projects in the developing world, who then sell them on to companies in the industrialised world, who need to add to their carbon quota. In effect it is an important means of industry financing carbon reducing projects and is a rare example of a form of global governance despite all its well documented shortcomings.

Possibly the single biggest incentive for governments to actually take action is the fear of the lights going out. Energy security, whilst not necessarily directly related to the climate change issue, is its best bedfellow. Whilst at present nations are happy to depend on other countries for the provision of part of their food supply, they have become increasingly concerned with the risks associated with the importation of energy. The Americans see much of their oil still passing through the extremely narrow Straits of Hermos each day with Iran controlling the northern bank. Western Europe is equally conscious that Russian gas has a long journey through several countries with taps all along the route, which can be politically turned off. Then the world watches as the Middle East ferments as the Israeli Palestinian issue smoulders, Iran moves to becoming a nuclear power and foreign troops fight in Iraq and Afghanistan, which itself boundaries on the less than stable nuclear Pakistan. Add to that the turmoil and inevitable struggle for power in the countries of the Arab Spring and just one spark is required to set off an inferno that denies the great modern day civilisations their fuel.

In the mean time oil and gas moves relatively freely round the globe by ship or pipeline. However, that freedom is also matched by its vulnerability of supply from the threat of, or actual, disruption of this flow by terrorist groups, hostile governments, increased home demand in source countries or market shifts. The need to increase energy self-sufficiency is widely recognised by just about all states and from an environmental perspective this has opened the door to the generation of greater quantities of energy from non-fossil fuel sources. It would seem that fuel security is coming through initially as the biggest force for change.

Whilst the role of government, be it in halting deforestation, providing green tax incentives, penalising extravagant use of carbon, supporting fledgling renewable projects or regulating individuals and corporations, cannot solve all the issues, it can play a key part in the decarbonising of society. Corporations in particular, and often against their natural instincts, are looking to government to set out a road map and provide a lead that enables them to factor in green initiatives, and regulation that provides a commonality for both them and their competitors. Leadership on climate change by governments is sorely needed and they have at their disposal the tools of taxation, subsidy, regulation and treaty; these, whilst limited, are potentially extremely powerful. It seems the tool box is full except for a hammer.

Corporations fit to jump

The role of corporations has largely been defined as 'generator'; they are portrayed as culprit. This is true for it is companies that use energy and emit greenhouse gases in order to service us with the essentials and non-essentials of life in support of the life-styles we choose to adopt. Where better then, than to turn to those who are culpable, for at least part of the problem, in order to find a solution? In so doing there is no value or purpose in taking the quick route and judgmentally decreeing that all would be well if these companies stopped their carbon emissions. The consequences of that would be dire and as with all things related to human carbon production we should aim to move quickly to a better place rather seek to make arbitrary and judgmental decisions.

Industry has for a number of years factored climate change and green issues in to their business plans. Every business needs to survive both in the short and long term and to do this they must meet customer expectations. Such expectations may be for more environmentally friendly goods and certainly for items that are competitively priced, as well as meeting the new perceived needs in a rapidly changing society. With energy prices rocketing all companies are searching out means of reducing consumption and some with green biased customers are moving to renewable energy.

A recent report on Samsung in South Korea highlights exactly how embedded such thinking can be within the corporate body. In that year Samsung sold $168 billion of goods and employed over 275,000 men and women. Such a huge business has enormous power both in terms of purchasing materials and in influencing the market place. But that power comes from size and with size comes problems of control and direction, as has been exemplified by Toyota and General Motors. Companies such as Samsung, therefore, have to think through their future very carefully indeed, for if they invest in the wrong market then the results can be commercially catastrophic; of course if they get it right they make a lot of money. Samsung is therefore playing for high stakes and what they decide is a pretty good indicator of what some of the world's best commercial brains believe to be the future.

In 2010 the group announced a $21 billion investment over the next 10 years into renewable energy and healthcare and in so doing diversifying from their established business as the world's largest manufacturer of electronic goods including mobile phones and computer chips. In other words they are predicting an ageing and more affluent population with health care needs and hungry for energy. The choice of renewable energy recognises that fossil fuels are dwindling and are of the past with security, climate change and pollution driving green energy in the form of solar cells, rechargeable hybrid car batteries and LED (light emitted diode) technology. As a post-script to the announcement the CEO noted that most of the group's current range of products and the businesses they support within the group are likely to have disappeared within a ten year time frame. This highlights the rate at which change can take place driven by an instinct to survive and to prosper; a very human response.

As a footnote, corporations do not always get their act together quite as well as Samsung. The state owned electricity generator and distributor in Nigeria is called the Power Holding Company of Nigeria (PHCN), which apparently is also known as 'Please Have Candles Nearby'. Its reputation is apparently well earned for whilst Nigeria, with its 150m population, is Africa's biggest oil and gas exporter, this company provides only enough electricity to run a refrigerator for one in every thirty people.

Originally most companies tended to be family businesses completely controlled by their shareholders. As these companies merged or grew in response to greater market access to potential customers, ownership took on a wider meaning and larger companies moved into public ownership. Today ultimate shareholder control remains technically in place, but the reality is that the shareholder has little influence over the business and the power of ownership has moved over to the customer. It is this disparate collection of individuals, or 'small people' as the chairman of BP referred to them when trying to contain the consequences of the Mexican Gulf oil spillage in 2010, that hold the reins and make companies jump; and whatever the size or reach of a company they will all jump to the consumers' tune.

Individual – the power house of change

I was a little surprised when talking with someone, who had adopted a very green lifestyle avoiding flying, using public transport and bicycle, recycling and so on. Whilst he had wholeheartedly adopted this means of existence he felt at the same time that it was a largely futile gesture and argued that only governments and large corporations could in reality make any real difference and thereby seriously diminish the rate of temperature rise. So, whilst his conscience dictated his actions, his intellect condemned such steps as being of little or no value. It was as if the scale of the problem was of such enormity that individual effort was deemed of little worth like whistling in to the wind.

As we struggle through a series of major financial crises and the world's financial power centre moves from West to East, Alan Greenspan, the former Federal Reserve chairman, then arguable the most powerful financial man on the earth, admitted that such was the demand for assets that all his efforts to control monetary flow had been in truth to little effect. The demand for assets of course comes from the individual, who wants to own a home and have a pension at the end of his or her working life and who are in fact, perhaps in absentia, cheering on the more adventurous hedge fund manager or private equity investor. In other words the behaviour of the individual and the large financial institute and bank are inextricably linked and identical.

However much we might protest about the behaviour of these institutions and rail against their unethical ways, we are quick to move our savings to a place of higher return or transfer our mortgage to a better provider. We are all greedy even if we mitigate this by using a fig leaf of claimed ignorance about financial matters. As for finance, the same applies to climate change; where the behaviour of the individual is paramount and is inseparable from the institutions, governments and corporations.

In essence it is the individual who holds the whip hand. The individual is the consumer, the individual is the voter or citizen and both governments and corporations, with all their facades of power, fear them and know they ignore them at their peril. This has never been more so than today with the emergence of instant internet and mobile communications. Just as the butterfly in South America flaps its wings and causes an earthquake in Japan so just a few words on Twitter can bring low what was once seen as an impregnable government or corporation. Chaos Theory is now an actuality held in the palm of everyone's hand. Those who doubt that should ask the Chinese ruling party, or Toyota or Nestle.

The question for us is how do those concerned about climate change spur the individual to act and in so doing generate governmental and corporate action. Part of the problem is a degree of confusion over whether we should be addressing the prevention of climate change or focussing on dealing with the consequences of such change. In truth we need to deal with both for it will take time to dampen the fires we have already lit and it is therefore inevitable that there will be consequences which we will have to address. In so doing people's perceptions and personal response have to change markedly.

In a study workshop entitled *East of England* the working group concluded that to move to a 60 percent reduction in carbon emissions by 2050 would require a huge shock perhaps resulting directly from climate change such as a series of heat-waves, or flooding as a result of excessive rain or sea-level rises, and/or fuel price increases driven by shortage resulting from conflict or diminishing supplies. A contributed paper from the Tyndall Centre estimated that if we just carried on as we are and continued to reduce our carbon dependency at the current rate then emissions would fall by 20% by 2050. However, if following shock therapy the market intervened, then emissions would fall by 60 percent.

Interestingly, for those who may wish to push a particular political agenda, replacing a market fix with government regulation or a society move to 'local and green' the 60 percent reduction remained unaltered. Politics, therefore, is not an accelerator pedal that will change the speed of change, neither are market forces or movements per se. What makes the difference is the individual and the individual needs a wake up call.

Shock therapy

From this we can only conclude that change will be slow, albeit progressive in our dealing with climate change issues until such moment when we are directly affected by an economic or physical crisis clearly seen to have been derived from climate change. That fits with the human character and is the repeated story of civilisations. Those who enjoyed the fertile soils, the benefits of sea trade and the affluent lifestyle of the city of Pompeii on the lower slopes of Mount Vesuvius, knew of its rumblings but ignored them until it was too late and they were overcome as it erupted for two days in 79AD. We, who live in the UK with its favourable location and historical store of affluence, also hear the rumblings. True, we are not higher up the mountain with those who are the first to suffer the full force of these shock waves, but for all of us it does not take long for the rumble to become an eruption.

Without doubt the biggest single effect of climate change will be on the ability of land to produce food, but if we are sadly reliant on moments of shock in order to educate people then it is more likely to be the floods, mudslides, typhoons and hurricanes which have always been around, but are now occurring at a much greater frequency as the temperature rises, that will get us off our seat. The United Nations have estimated that in the next five years an additional 100 million people will require their help as a direct result of an increase in the number of natural disasters caused by climate change. This is a 50 percent increase in such occurrences in 1800 days. It seems we must as always learn the hard way.

We all live in a world that is experiencing temperature increase and we all know and acknowledge that. We debate a little about the likely consequences, but most of us as individuals would agree that the prognosis is not good for us or our children. Around the world it is the individual who votes for a manifesto and elects a government or takes to the street and creates change, often at great personal cost. It is individuals who choose to buy a company's goods or walk away from them. It is each individual who holds these powers of censure and each individual who can demand change. In other words each of us has the power to take the necessary steps to stop or mitigate climate change. All we need is a reason to start on such a course. Some have taken the first steps already and through their experimentation we are leaning much that will undoubtedly prove essential in meeting this challenge. Others, the majority, will only join them when they are shocked into action. Whether it is then too late or not, only time will tell, but it is only then that things will happen at an effective and accelerated rate.

If the history of humankind is one day written by a visiting alien, it is inevitable that the tome will contain much discussion on why a planet so fabulously rich in energy delivered round the clock from outer space, both gratis and safe, should have become so reliant on using pollutants to power their needs; pollutants that warmed the atmosphere and eventually precipitated their extinction.

EVERY CRUMB COUNTS

The problem in the wings

In the late 1970s I gave a short paper relating to the question of feeding the world as the population swelled. It dealt with the same fundamental issue we now face albeit it was framed in the context of knowledge of a world some thirty years younger. It is that time distance that makes the contents, or at least parts of them, of some interest;

Farming was born out of the need of man to feed. The long and precarious story of primitive agriculture was based on the concept of personal and family survival. To farm was to deal with a matter of life and death. Consequently individuals would have been cautious and by trial and error, often costly, they established safe tenets of husbandry. Thus they ensured, to the best of their ability, the future of their own food supply. It is only relatively recently, following the Industrial Revolution, that food production has increasingly been considered as primarily a commercial, profit making industry. Even this new found role of farming for profit was tempered by the traditional husbandry safeguards. With the development of the consumer society and the encouragement of the critical faculty, it is hardly surprising that over the last few decades agriculture has broken free from rotational and other husbandry constraints and the farming business, like those of its urban cousins, has been based on the maximisation of profits. But where do we go to now? As the oldest industry, where is our new maturity going to direct us? Is farming for the maximisation of profits compatible with the real world situation and the need to maximise food production?

If food is to be produced in the most efficient way in food value terms, then farming must shortly take on the mantle of a social service. As a nation we have accepted the right of every individual to receive medical treatment in order to sustain good health. This right to good health is a social service embodied in the NHS. Again as a nation we have accepted the principle of equal opportunity in education. This too has become a social service carried out by the state. If we accept the right of the individual to have provision made for good health and education, then it is an easy step to the acceptance of the provision of food as a social service. As food producers, we as farmers now face the biggest issue history has ever offered anyone. World conditions have placed us in the major role demanding considerable wisdom and responsibility. We must now put agriculture on a new footing, a war footing, to tackle the question of survival.

Talking to a farming audience it was not well received! Since then, when 1972 had seen the first fall in world food production since the Second World War and the seemingly invulnerable green revolution lay damaged, the globe's food producers have coped remarkably well with a population that has grown by 50 percent. The buffers were hit again in 2007 when supply fell back as weather conditions and the increasing uptake of cereal crops for animal feed and biofuel denied people food. It is inevitable that those buffers will be tested increasingly frequently in the years to come and forecasted end of season stocks in 2012 already look scarily low. It seems as if the search for food has once again been galvanised and the issue has become a matter of enough importance to merit some newsprint. Nevertheless, sometimes is does appear that only a very few have really grasped the magnitude of the matter and the problems that lie ahead, barely concealed and waiting in the wings ready to ambush mankind.

Agriculture's pedigree

It was approximately 30,000 years ago that hunter gatherers first began to settle and to start to farm. Over the many thousands of years that were to follow naturally occurring grasses were bred into cereal crops and wild animals domesticated to give milk, eggs, meat and wool. These activities were the bedrock of civilisations and without them societies with their towns and cities, education and culture would not have formed. Economic take off in the West commenced with wool and its early trading, which created wealth and all the benefits that wealth brings.

Early paintings on cave walls and the beautified stone carvings and sculptures of the early civilisations of Egypt, Phoenicia, and Syria all bear testament to the understanding these people had about agriculture and its importance to survival and wellbeing. We are now so remote from such matters that growing food seems to have little relevance to life and its living and yet that ancient wisdom and understanding is still held by the Maori or the Masai.

In the UK the last 2000 years has seen a steady progression in farming techniques that trickled through at first before becoming a torrent of innovation in the last century and into the new. One of the earliest advances was the introduction of oxen as draft animals to pull carts and the all important plough as it inverted the soil and improved aeration, soil fertility and weed control. Yields improved and such was the value of these animals of burden that they would be housed in the peasant's own house with the added benefit of their then having a live live-in central heating system.

In the UK, as society evolved with its defined class system, the landed gentry turned their minds from warring to improving the breeds of cattle, sheep, pigs, and horses, which took over from the oxen as the main beast of burden speeding up operations and introducing greater flexibility. Farming was no longer the sole domain of the underclass and its importance in society became increasingly evident.

Thus the Agricultural Revolution began in the late 1600s and was to last for over a century. It was this revolution that supported and underpinned the Industrial Revolution and the urbanisation of the population. Land was enclosed into the field system we are familiar with today enabling farmers to improve their management of crops and stock simply by controlling what went on or how a particular parcel of land was treated. The Three Field Rotation was introduced whereby farmers grew wheat, then barley followed by a legume crop in succeeding years. The legumes replaced the old non-productive fallow system and with its ability to fix atmospheric nitrogen via root nodules they greatly improved soil fertility. Eventually these early rotational systems gave way to the more sophisticated Norfolk Four Course Rotation which included animal grazing to further improve soil fertility and structure.

At the turn of 18th Century mechanisation began to creep in and Jethro Tull introduced his seed drill greatly improving germination, and with the cereal plants set in rows, it allowed for mechanical hoeing and the elimination of weeds. It was also through this period that animal breeding became more of a science and many of the forerunners of today's breeds were developed.

With industrialisation in the 19th century mechanical means of doing jobs that had always been done by hand were introduced at an accelerated rate. This included the harvesting of corn and its subsequent threshing, which caused some unrest in the countryside as jobs were lost. Those like the Tollpuddle martyrs revolted against such change and in so doing gained their moment of history before being pushed aside by the inevitability of progress.

Whilst the great driver of the industrial revolution, steam, was used to a limited extent on farms it was not until the introduction of the internal combustion engine that a second wave of change was experienced and almost overnight the horses were slaughtered and replaced by the tractor. All this took place against the backdrop of The Second World War and the urgent need for home produced food. With the introduction of greater power the door was opened to cultivate land that had been too heavy for the horse to turn and to improve crop husbandry further. Consequently in the second half of the last century science dominated farming developments with chemical fertilisers, herbicides, pesticides and accelerated breeding programmes. For a period yields rose almost without limit and a starving Europe was transformed in a matter of three short decades and became a place of overproduction, with milk and wine lakes, butter mountains, and barns filled to the brim with corn.

Farming science

Science, of whatever kind, presents us with a Pandora's box. The benefits are massive to health, wellbeing and enjoyment and yet each time it seems there is a reverse side to the coin, and we are faced with political, social and ethical dilemmas. The splitting of the atom has delivered clean energy, but also the nuclear bomb and the means of destruction. The development of the internal combustion engine has granted mankind mobility and a release from toil, but has also polluted and added to global temperature increase, and genetically modified crops have significantly advanced yields whilst in some places leaving a trail of despair and concerns over a narrowing of the gene pool. How we balance the benefits and dangers that scientific discovery brings will be a demand placed upon mankind at an ever increasing rate over the years to come and how successful we are dealing with that will predetermine our ability to survive.

Today agriculture is engaged in a whole series of scientific developments, some of which are large scale experiments that will undoubtedly have profound consequences on what we eat and indeed whether we can eat. Genetic manipulation is one of the prime drivers and the industry is offered huge rewards for the adoption of these products and techniques in terms of improved and more secure yields whether it be the growing of crops or animals. The fact that thousands of small farmers in Southern India have committed suicide caused by their being persuaded to use GM cotton seed may be argued by some as an inevitable and an unfortunate part of the change process and that the long term benefits outweigh any such human misfortune. The fact is that GM cotton seed costs considerably more and requires supporting husbandry techniques of using fertilisers and chemical spray regimes in order for the crop to realise its much improved yield. This works well and the logic of the farmer borrowing to finance the costs of these seeds and chemicals seems sound until the weather causes crop failure. In times past the loss of a low cost crop could be survived, but now with high inputs of borrowed money there seems no way out and a man who cannot support his family has no purpose and chooses to die.

There are other examples of GM crops resistant to insect damage causing a population explosion of these same insects at the margins of their production resulting in immense damage to crops on adjoining lands mitigating the overall beneficial effects. And of course there are those who argue with some intellectual integrity that GM will inevitably reduce the gene pool as each crop type is dominated by a single or small handful of GM varieties.

In the 1980s when I was directly engaged in commercial growing there was a degree of mystification amongst the farming fraternity as to why the oil and chemical companies were buying up seed producers. They seemed poor bed fellows with the chemical and oil giants armed to the teeth with sophisticated technology and assets, whilst the seed companies were small and unsophisticated, working year in and year out to develop new varieties primarily using proven and traditional propagation techniques. What was then an odd match now makes eminent business sense as companies like Monsanto sell their chemical herbicide Roundup, which is now the biggest selling herbicide in North America, as part of a package with Roundup Ready seeds. In other words, GM modified soya bean seed can be sown that is resistant to glysophate, the active ingredient of Roundup, in the knowledge that the resulting crop can then be sprayed with this systemic herbicide and all the weeds will be killed leaving a crop undamaged and without competition and with the potential of high yield.

Enthusiasts argue, with credibility, that in the context of oil based farming practice this method reduces the number of tractor passes, minimises chemical inputs and improves productivity or put another way reduces the carbon foot print of each kilo of soya beans produced. The problem is that Monsanto hold the controls and whilst they claim that these levers are used responsibly and their products are essential, if we are to feed a burgeoning world population, these levers are nevertheless beyond the reach of the consumer, the processor or the farmer.

The dilemma is borne between maximising yields per hectare to feed a growing population and the risk that such technology will leave us open, through its narrow specificity, to collapse. How much do we risk to feed the many when failure of these techniques would feed only the few and result in an even greater catastrophe?

In Europe GM crops remain banned save for a license for Spain to grow a relatively modest acreage of GM maize. Beyond the Continent things are very different and in North America, South America and the Indian

Subcontinent huge acreages of GM crops are grown as part of the year in and year out cycle of crop production. In the early 2000s I overheard a man from Nestle propounding the theory that European governmental legalisation banning GM crops was irrelevant as they would contaminate and overrun the food supply system irrespective of laws. To a certain extent he is already right and in the UK we regularly consume GM soya bean products solely because there is little other choice.

In 2009 The Royal Society entered the debate about GM science with its report entitled *Reaping the Benefits: Towards a Sustainable Intensification of Global Agriculture*. Its primary purpose was to ascertain biological approaches to enhance food-crop production with the objective of raising world crop yields by between 50 percent and 100 percent. There is a refreshing degree of clarity of thought in this report and whilst it gives a nod to non-scientific issues, the focus on, and hence the cause of that clarity, is scientific solution. This is science at its best irrespective of whether or not such issues as GM are considered as positive or negative.

Taking their thoughts on further if, for example, we could genetically engineer the characteristic of legumes to fix atmospheric nitrogen and imbed it in the wheat genome then the benefits could be quite remarkable. The need for artificial nitrogen fertiliser would be no more, delivering massive energy savings, reducing waste through soil leaching and as a result reducing stream and pond eutrophication. Equally the sudden and worrying re-emergence of wheat rust that once devastated crops causing great hardship will do the same again unless checked. This may well require the quick fix of genetic modification to hamper its spread and once more defeat it. Whether we intuitively like the new science or not, we have a duty and obligation to at least listen to the case and open our minds to it, for each new idea could influence the lives and survival of the many.

Marketing science

In the 1980s I was a regular visitor to California and the iceberg lettuce growing areas around Steinbeck's Salinas and the Mexican border lands around Yuma, Arizona. The purpose was simply to understand the techniques and methods of iceberg lettuce production and to transfer these to the UK and Spain. The industry in this Western State was advanced and companies would grow 20,000 plus acres of lettuce using the length of California to deliver year round supply. Much of the lettuce would then be shipped by truck to the East Coast often by married couples, who shared the wheel and drove non-stop from one side of America to the other.

The growing process was sophisticated and very American, with the land first being levelled using huge mechanical planes fitted with laser devices. The soil was then raised into beds upon which small lettuce plants would be mechanically planted out, straight and perfectly spaced. Irrigation was by inundation and the pumped water flooded down the rows between the ridged plants ensuring the leaves remained dry and the roots well watered. This form of flood irrigation demanded feet of water, whilst back in the UK we were used to applying just a few inches. With full and programmed fertiliser and chemical treatments the fields would come to harvest within 120 days; when the lettuce heart was deemed not too soft and not too hard by a well-practiced hand and the window for cutting was two or at most three days.

This required self-propelled harvesting rigs slowly moving across the field covering a span of rows each with a 'cutter' walking along their furrow and cutting and placing the heads on a moving belt or static table from where a 'wrapper' would take the head and wrap it in film before passing it to a grading table where it was boxed according to size. Field lorries would drive alongside the rig and the boxes were unloaded and taken to a central vacuum cooler where within two hours of cutting, the lettuce was cooled to 2°C and stored under refrigeration ready for that day's transportation. We would base our stay on Salinas as that was the headquarters of this industry, however, because our visits tended to be in the winter months, we would be flown down in the company jet to the southern most harvesting areas right on the Mexican border.

The American 'can do' and 'open door' attitude was an eye opener to me, having at the time only experienced a more stodgy UK management style. It was refreshing, invigorating and exciting and those who ran these companies enjoyed and shared their success, which seemed to be conditional on their having had heart surgery and to have gone through bankruptcy at least once and preferably twice on the way up.

At the time of my early visits everyone was celebrating the arrival of a variety called *Salinas*. It was a high yielding, visually appealing and a robust variety that was relatively easy to grow; the American dream lettuce. The man who had developed this strain was part of a government funded service and when I met him he was a quiet, considerate and enthused guy who worked out of some rather old, just functional greenhouses and offices. Nevertheless he was lauded by the whole industry, was a well known name, and was guest speaker at conferences and meetings across the State and beyond. The commercial yields of lettuce jumped as the planted acreage of the *Salinas* variety increased almost exponentially.

When I met him eighteen months later, I was taken aback to find a rather subdued and introspective individual. The reason soon became clear and as we sat talking over a large cup of coffee. He explained that his variety *Salinas*, recently acclaimed by growers as an industry lifesaver, was now considered by the same people as having ruined their businesses. His reception at gatherings of growers and traders had become hostile and unpleasant for, as he explained, the yields had gone up and the market's response, as always to surplus supply, was a price crash. He was deemed the cause of this misfortune and yet all he had done was to give the industry a better product.

This was a classic case of science leading the way and becoming out of sync with the market or the needs of people. Whilst this was perhaps a blip in the history of one particular crop, which is probably long forgotten, it does highlight the need for science to work with the other disciplines of life rather than just independently. Science is a little bit like water that always finds the easiest route to a new level, for researchers tend to follow their own train of thought in a creative rather functional manner, in preference to understanding the full needs of the society first and then seeking to respond with their science.

The early scientists were indeed philosophers, thinkers, theologians and their science was often just an extension of their thoughts and faith. This made them more rounded men, who tended to look at life as a whole and their science undoubtedly benefited. Since then science has magnified knowledge like one of its great telescopes, but for the individual scientist the demand has been to peer down the microscope and see smaller and smaller fragments to be researched. This makes it very easy for the researcher to become remote from society working on things that are not understood or not perceived to be relevant. Such criticism is in part fair and the need for society to work with the scientist and to seek out projects and matters that are urgent and necessary for the survival of our biosphere and our own species within it, has never been greater. It follows that we must ensure that the greater proportion of science is applied research and well applied.

Agriculture and The Question

The key question we have set ourselves is whether we can sustainably feed a population set to grow to 9 billion plus by 2050?

I attended a symposium on exactly that subject with a well qualified panel including representation of farmers, manufacturers, retailers and NGOs as well as a think tank. The food crisis of 2007 and the market's response was still on their minds albeit their analysis and understanding of the consequences was heavily coloured by their particular interest. The fact they all agreed upon was that food production would have to increase by at least 70 per cent in a matter of 40 years. The Oxfam man highlighted the plight of sub-Saharan Africa and the fact that globally the 800 million who went hungry each day in 2006 had now increased to 1.1 billion. The 2006 crisis was in staple foods such as rice and as a consequence, 34 countries including India, had banned exports in order to preserve food stocks for their domestic consumers. This not unreasonable move had immediately spooked the market and prices had increased dramatically. Whilst this was deplored by the manufacturer, who was quite reasonably trying to control costs, no recognition was given to the adverse effects on those less economically developed nations, of government subsidies to growers in the European Community and USA to keep their prices artificially low, or the fact that an ever increasing share of the cereal crop is now being diverted to biofuel production.

The National Farmers Union leader felt that self-sufficiency in food was achievable in the UK, although he considered it undesirable on the grounds that we should grow those crops where we had a natural advantage and trade surpluses of them for other more exotic produce. He went on to note that the large tonnage of protein we import from Brazil and other countries for animal feed was more vulnerable to changes in supply. Even the Oxfam representative expressed the hope that food trade would continue as he considered it a net benefit to the poorer countries in which Oxfam was engaged. All agreed that science and the technology it generates were key to feeding the increasing number of mouths. Two examples of the successful application of GM technology were cited. For the UK farmer work at the John Innes

Centre to develop a GM blight resistant potato would have a very positive benefit and for the Bangladeshi farmer GM saline tolerant rice was already proving itself of immense value on lands frequently inundated by or subject to aquifer and sea water ingression. And so the discussion circulated the forum with mention of waste and then finally climate change almost as an afterthought.

What was evident was that the manufacturer saw the problem of food delivery as solely one of cost, the Oxfam man interpreted it on behalf of the poorest of the poor, the farmer's view was one of opportunity for his community to receive improved returns, the think tank woman treated it as almost an academic issue and the retailer as one of little consequence. Each member was on top of his bit of the subject, but none could or seemed to want to see the whole picture. Perhaps the farmer had the last say for when pressed on his industry's ability to supply in the face of climate change he confessed, 'the next 10 years are ok, but after that climate change may well make everything very different'.

In the 2007 crisis when the cost of grains rose alarmingly, what became clear was that those crops that could be used for biofuels and therefore have a dual demand, experienced a significantly higher price hike than their 'food only' counterparts. Today's market is underpinned by the need for food and driven by the new energy rush for ethanol.

A common theme throughout all such discussions is the need to liberalise the global trade of food stuffs. The distorted market of recent years has meant that highly subsidised agricultural commodities from the industrialized world have undercut those same commodities produced in less developed and more agrarian economies. This has meant that, despite their natural propensity to produce a particular range of crops, their potential strength and place in the world market has been neutered by foreign subsidised agriculture, and their farming sector, which should be thriving, remains suppressed and without investment income.

Once the world's farming and associated trading is freed from such manipulation then there can be a free flow of goods and the tendency to produce in the most naturally advantageous areas will once again come through.

Small barns

The world only holds a very limited stock of food at any one time. The majority of stored foods like wheat, soya, rice and other grains are grown in the Northern Hemisphere with its harvest stretching between May and October. In effect the world's warehouse begins to empty through the rest of the year and it is not unusual for it to hold no more than two months' supply at the point of carry over to the next season. In 2009, with a sequence of difficult weather patterns the USDA reported that at one point the States only held 14 days supply of soya beans, whereas since then, with a much improved sequence in the world's weather together with the impetus of higher prices, yields have increased and the stores have been refilled. Nevertheless for an essential that takes anything up to three hundred days to grow from seeding to harvest this seems a scary level of stock to hold; it is not even operating on a just in time cycle.

It is incredibly difficult to forecast precisely the future uptake of such stored crops where a whole range of factors can impact on demand. In any one year weather variation alone can affect global yields by anything up to 4 percent. Add to this regional crop failure due to disease, exceptional cold spells increasing the level of animal feed required, or new biofuel processing facilities coming on stream with their almost insatiable demand for grain and a very uncertain picture emerges. Not least is the ability of the farmer to up and down his game. The trouble is that he is always a season behind the market.

In 1976, when we experienced an extremely dry and hot summer following a winter of very low rainfall in the UK, I was growing 500 acres of Brussels Sprouts. They were mainly planted on fertile Brickearth soils, but were beyond the irrigation network which was anyway at full capacity working night and day on salad crops. The plants grew slowly, hungry for water, whilst the high temperatures suppressed vegetative development. Knee, rather than waist, high the fields of sprouts looked poor and when the rains finally came in early October, it was too late to allow for anything other than a final short spurt of growth. Still in an age when the produce markets were largely reliant on home grown vegetables there was only one outcome and that was a low yield and very high prices. Indeed, even then we were receiving a price that equated to 3p per sprout and the profits rolled in. Of course the following year, on the news of such returns, growers squeezed in an extra few acres here and there and after a more normal growing season the market in the following winter collapsed as a result of abundance of crop and the permanent loss of some customers who had moved on to the alternative vegetables they had tried during the high price period.

The lesson remains that given enough reasons, farmers and growers will magically conjure up acreage and increase productivity. This was brought home to me when I moved up to South Lincolnshire, which holds some of the best soils and has always been at the forefront of British farming. Initially I saw the sort of crops I had expected that were well husbanded, even, weed free and vigorous, but as the wheat and barley prices fell and set aside took hold the only response was to reduce costs. There then appeared to be a loss of interest and pride with the consequences the fields looked unkempt and poorly farmed. However, since 2007 and the food crisis, prices improved aided by the additional crop uptake for biofuels, and once again the fields looked well farmed with high yielding crops.

This sort of flexibility and short response time has been the hidden strength of the farming community and has enabled it to prosper in good times and survive, albeit with hardship, in harsher economic climates. The farmhouse we lived in during our time in Kent bore witness to exactly that, with the original dwelling having been built in the 17th Century with an extension to the front and rear of the house. The front extension was quite grand and had been erected during the good years of the Napoleonic Wars when grain prices hit a high, and the rear addition was of Victorian design likewise built when cereal prices had move up as a result of the Corn Laws.

The Soil Association, like most campaigning groups, is a mix of credible facts and campaigning zeal, which creates its own precarious but important balance. They have questioned whether food production in fact needs to rise the widely reported 50 per cent by 2030 and doubled by 2050 in response to world population changes. They suggest that this figure is exaggerated and instead of reading 100 percent by 2050 it should be 70 percent. Of course both figures are hugely challenging and perhaps at this stage debating the difference is a little semantic albeit the difference in real terms is indeed considerable. Their line of argument rests on questioning some of the key points in these UN projections. The Soil Association argues that these projections incorrectly assume that there will be a global move to a Western 'unhealthy' diet, that food security is more than the simple matter of growing enough food as measured in the UN studies and must include its accessibility, and that the assumption that current mechanisms and flows of food trade remain unaltered is not a given.

One of the consequences of the UN assumptions is that much of the extra cereal production will in fact go to feed the extra 1 billion cattle that they predict will be required by 2050 to feed our demand for meat and dairy products assuming the widespread transference to a Western style diet. The Soil Association concludes that such a scenario would exacerbate the problem of GHG emissions massively with this number of cattle producing catastrophic volumes of methane. Their study concludes that the world can indeed feed 9 billion people using organic means of production provided 'modest diets are adopted'.

Eating habits, where food is plenteous, certainly do dictate the market. The little piece of ingenious research conducted at Cornell University on meal size over the centuries was a fascinating such insight. They simply looked at 52 paintings of the Last Supper, which had been painted over the centuries, and measured the size of plates, foods and bread and related them to the size of people's heads; the constant in the equation. They were able to argue quite convincingly that people tend to eat bigger portions on bigger plates and that this leads to an increased prevalence of obesity. Their findings concluded that meal size had grown by 69 percent over the past thousand years. Today America owns up to having 31 percent of its population as being obese, which supports these findings, and when travelling in that country one soon learns to ask for a small portion of everything.

The Role of Livestock

My enjoyment of farming animals has been immense. When I was at primary school I went on holiday in a parked caravan on a farm near Exmouth in Devon. I watched cows being milked by hand and was given a ride on a horse drawn cart as it plied its way there and back between farm and potato field. Maybe that is what triggered my love affair, for if you can fall in love with cows then I did as a seventeen year old experiencing farming for the first time in North Yorkshire.

I had lived in the same house in Beckenham in the south suburbs of London all my remembered life and approaching the end of my school years I sought to join the farming fraternity. I knew nothing and knew nobody. Fortunately my mother recalled a girl she had known in the war years when she had been stationed by The Salvation Army in Attercliffe, who had lived in the slum area of the town near the steel works and who had joined the Land Army and then married the farmer to whom she had been assigned. He tenant farmed 120 acres at Barnaby Sleights on the hills above Whitby and so letters were exchanged and I was duly dispatched to North Yorkshire to experience farm life for three weeks during my summer holidays. I journeyed up by steam train, changing at York, before eventually arriving at the quayside station in Whitby. There I was met by the farmer and his wife and we set off for the farm.

Every moment was a revelation as we weaved and bumped our way through narrow lanes and then up the farm track well rutted by rain and use. The farmhouse was set on one side of a farmyard with stock sheds and barn completing the square with the lavatory, an earth closet, set at the far side of the vegetable garden that stretched out in front of the house; in hindsight it was even then something out of much earlier times.

The farmer, Albert, was older than his wife who busied herself inside and out undertaking everything with a degree of relish and cheerfulness. The eldest of three daughters helped on the farm, whilst the other two took the bus to work as shop assistants in Whitby. The cows all had horns and came in for milking to stand in a small shippin, or line of cow stalls, waiting to be tethered by neck chains whilst the twice daily ritual of milking was conducted with two rather basic Fullwood milking units linked to a simple air line. The milk was tipped into a funnel that cascaded the white liquid over a water cooled surface before depositing it into a churn.

The Friesian bull lived tethered in a dark stall, which was only just big enough to contain his huge form and when I was instructed to feed him with hay, I had to push between him and the once whitewashed wall to get to his manger, providing me with moments of both trepidation and accomplishment. When a cow was on heat the animal did not return to the meadow after milking but was left in the yard to await the rather clumsy service of the bull freed for a moment from his dark and solitary existence. A few years later I was to read the opening paragraphs of Emile Zola's book *Earth* and understood precisely the scene he described. Why this world of antiquity, earthiness and hard work appealed to me I do not know, but I savoured every minute and fell in love with cows.

In contrast in 2010 Nocton Dairies floated an ambitious scheme to house 8000 dairy cows near Lincoln. The concept was very modern with on-site veterinary attendance at all times, and with the cows housed on sawdust and kept immaculately clean so that their life of giving milk would be healthy and sanitised. The only time they would be out to grass was whilst dry awaiting the cycle of birth and lactation to start yet again. The collected slurry was to be used as a biofuel to power the plant and on paper the argument ran that the animal's welfare would be paramount with control over disease, nutritional needs and the provision of a safe environment.

The company withdrew their planning application in the face of considerable local opposition possibly more concerned about light pollution, round the clock activity, an element of immigrant labour and a stream of lorries collecting the milk rather than animal welfare matters. But this concept, whilst well thought through, was centuries away from Zola or my experience at Barnaby Sleights.

The farming fraternity is a marvellous community inevitably representing a mix of society but to a degree tempered by their interface with nature and its forces. It is a fraternity that is changing fast as it ages and industrialisation penetrates its practices. Today the average age of Uk farmer is a few months shy of 60 years. Within a single community it is indeed difficult to accommodate the dairy farmer, who milks his herd twice a day, seven days a week, fifty-two weeks a year, with the owner of a super dairy farm housing thousands of cows indoors in a million square foot unit year round. The former is going out of business as the latter moves in with all their claimed efficiencies. Prima facie the stock will be protected from the badger, TB, foot & mouth, and the unit will make use of round the clock milking and ease of collection, but what has the industry achieved at the end of the day?

Follow the history of the poultry industry; the word farming has been dropped. Here there has been a move to intensively caged birds whose sole purpose is to produce eggs or meat on as meagre rations as possible and as quickly as possible. That is good practice in the industrial sector, but because a sentient animal is part of the process it has been found wanting and unacceptable to a vociferous section of consumers. Even in the farming world it is not readily or enthusiastically spoken of. The Northbourne question creeps in again – we abhor the process but do not speak out for all the reasons of denying people cheap food and fellow farmers a living.

The pig industry has followed suit and now we are well on to taking the bovine sector the same way.

In the 1980s and 90s from time to time I used to fly over the feedlots in America and observed the thousands upon thousands of beef animals closely housed in yards feeding their way to as an early death as was commercially possible. Once we stop sharing nature with those we farm we became a little less human.

Grazing animals on grassland seems somehow right. Where land can only support grass such as on the hill, marsh or river meadow the presence of cattle and sheep seems as in place as the grass itself. There is a harmony that lifts the soul and many such a vignette has been caught on artist's canvas, through the photo lens or expressed in music and words. No such expressions have been generated in the battery shed or feedlot. Once we enter the world of industrialised animal rearing we have to ask the fundamental question as to how it can be right to feed a beef animal 10 kilos of corn to produce 1 kilo of consumable meat? For pigs and poultry the conversion rate is less wasteful as the maintenance demand for smaller animals diminishes. Nevertheless however the issue is examined the feeding of people with cereal based foods is much more efficient than taking the route to feeding them via the animal. It is interesting that the great 'efficiency arguments' for intensive rearing are less well rehearsed when we come to our own species.

The arguments for meat production will inevitably become politically and morally less tenable as food becomes increasingly scarce and of higher price. Of course many consumers are hooked on their meat diet and many poorer people aspire to such a diet. That is leading the world through a period in which meat consumption is increasing massively. The consequences are simple – more animals, more carbon dioxide and methane, more deforestation, more food going to animals, staple food prices increasing and more people going hungry. Some would add more suffering, for the treatment of animals in the industrial sector of the old developed world has been dehumanized and in its own way is little better than some of the poorer practices and husbandry in the developing world of the East and South America.

But this growth in meat production can only be short lived. Put starkly, if the choice emerges between feeding a fellow human or a domesticated animal then meat no longer becomes an option. Of course the market can obfuscate such a pronounced choice and the rich can eat their meat whilst the poor die at their gate, but governments may well find this politically unacceptable as a destabilising force within their particular societies.

The world's per capita consumption of meat has increased from 30 kilos in 1980 to the present 41.2 kilos; a 37 percent increase. Meat consumption particularly has experienced rapid growth in those countries currently accumulating wealth and whose populations have also gained a taste for the Western diet, which remains heavily meat orientated. The economic downturn of 2008 and onwards has caused a corresponding downturn in the level of meat consumption. Meat is a relatively expensive food and it would seem that when money gets tight so the diet adapts. Even in the USA. with a second recession threatening, the rising cost of turkey meat in response to increased grain costs has severely supressed sales for the 2012 Thanksgiving.

These movements in demand complicate a market governed by the animal's cycle of birth and death. In the instance of beef, when demand picks up, then the farmer or rancher experiences higher prices and looks to expand his herd. This means that some of his heifers are retained as breeding stock and the period of shortage is extended and the shortfall is made temporarily more acute until those animals in turn produce calves that are reared and grown as beef animals. The whole process can take two or more years meaning that any response to movements in the market is a relatively long process.

The world population of bovines is 1.5 billion head and this is forecast by the FAO to increase to a level of 2.6 billion by 2050. That is an increase of over one billion cattle, all of whom will consume disproportionate amounts of precious grain and emit methane, a greenhouse gas that is eighty times more damaging in this regard than carbon dioxide. The population of poultry is similarly forecast to double in the same period. This growth in the number of domesticated cattle, pigs and birds means that of the estimated extra 1.3 billion tonnes of cereals required by the year 2050, some 550 million tonnes will be taken up by growth in the farm animal population, not to mention it increasing the chances of a serious avian or swine flu type virus that could mutate to humans and sweep the planet like an uncontrolled fire.

There is a school of thought that argues that the human diet needs to include high protein meat. What it overlooks is that for example one of our staple foods, bread, contains 10 percent of protein, which is more than enough for our needs. In the USA the consumption of protein is double that consumed on average in sub-Saharan Africa. That is perhaps not surprising in itself, but what is of greater interest is, that in the USA two thirds of the protein is derived from animals whilst in the sub-Saharan Africa region animals only contribute 20 per cent of consumed protein. Globally 37 per cent of consumed protein comes from animal sources; half comes from meat, with the balance coming from dairy and egg products. This immediately raises the fundamental question as to why we cannot at least halve our dependence on animals for protein and in so doing improve the environment by reducing greenhouse gas, reduce the pressure on cereal production, better human health through diet improvement and reduce the risk of animal derived viruses as exemplified by swine and avian flu, and allow for enhanced animal welfare. It should not be overlooked that of all the food groups animal products are by far the greatest contributor to greenhouse gas emissions with a Dutch study showing them to be responsible for 53 percent of the carbon dioxide released through food production.

For those who have worked with farm animals it will come as little surprise that research work carried out at both Bristol University and Cambridge University concluded that every animal we intend to eat or use is in fact a complex individual with remarkable cognitive abilities and capable of feeling strong emotions; they even worry about the future and get excited by solving intellectual challenges. The herdsman or shepherd knows only too well the social structure that a herd or flock create and how each individual animal has its own personal good and sometimes less good traits. Whilst modern husbandry does much to alleviate the symptoms of disease as well as care for the animal's physical wellbeing it increasingly crosses the boundary into what must be mentally stressful conditions by close proximity and indoor housing. As with people when the stockman can no longer know his charges by name then the relationship is impersonal and lost and individuals become nothing more than units of production.

In the same way that ceasing deforestation in order to reduce climate change seems obvious, very logical and do-able so the reduction in meat eating appears to be the most obvious and simple means of mankind addressing the global shortfall of food. Such a move would release grain and valuable protein for human consumption and reduce greenhouse gas emissions, so that in a stroke two of the great global problems are significantly addressed. Governments and corporations are powerless to stop such a change and the only possible impediment is the individual. A meatless diet may mean a change in food preference, but it is healthier, good for wellbeing and can be extremely enjoyable and in the kitchen less demanding.

The farmer redefined

Farmers are all producers of food, but their means of achieving this are very different when considered on the global stage. At one extreme there is the time honoured peasant, now virtually extinct in the West but very evident in the less developed countries, who scrapes a living from sometimes unyielding land assaulted by hostile weather patterns. At the other end of the scale there are farmers producing from huge acreages, who span the gap between field and plate by adding processing to their growing in order to add value.

The industry refers to the largest of these as 'ABCD' or Archer Daniel Midland, Bunge, Cargill, which are all American businesses plus France's Louis Defrus. Cargill is the largest with a turnover of nearly US$120 billion and remains a privately owned company now working in sixty eight countries with 137,000 employees. The business was founded back in the late 19th Century and has progressed from trading grain for the mid-West farmer to becoming the world's largest sugar and cocoa trader with heavy involvement in cotton and soya beans as well as cereals. It is a major supplier of animal feedstuffs and is directly involved in intensive poultry and pig production as well as biofuels, and natural replacement products for plastics.

It is easy to conclude that for such a large business life is plain sailing, but its current success has been borne out of generations of hard work, much risk taking and several near calamities as ventures and markets in Russia, South East Asia and most recently the food commodity price blip in 2008 ran them financially close to the wind. Currently they are working to a 2015 Strategy that is very aware of the power and importance of food production to a globe that can only become hungrier and more demanding of the Western diet. They are big, their intent is ambitious, but they are a long way away from our understanding of an empathetic farmer.

Whilst we are familiar with a relatively stable agricultural economy and community, in some parts of the world the demand for food and exports has driven massive changes to other parts of global agriculture and in quite short order. In just twenty years Brazil has powered its way through remarkable changes and is now a powerhouse of global farming driven by huge farms covering sometimes hundreds of thousands of acres in the high lands of Central Brazil. These areas have been developed with sophisticated international investment and unsophisticated chalk, which has been used to reduce the soil's acidity.

Investment has come with its own demands for proper labour laws, environmental awareness and protection and good relationships with indigenous peoples to such an extent that much of the lawless abuse of the land and the people who lived on it in this remote region has been banished. The net effect is that Brazil is now the world's largest exporter of sugar, green coffee, orange juice, chicken, beef, soya bean meal, ethanol, beans and oil and lies as fourth biggest exporter of pork and maize. No system is without fault lines and whether this apparent agricultural success story remains sustainable and environmentally worthy can only be tested by time, but it does present a good example of man's ability to turn on the tap when he has to.

In the poorer areas of the globe in Africa and South East Asia the opportunity to do a Brazil is less likely. Some argue that the 'land grab' by sovereign funds of more developed and well funded foreign countries has the benefit of introducing more advanced agricultural techniques which will spill out and extend beyond their boundaries to improve the lot of indigenous farmers. This acquisition of land is undertaken by states buying up tracks of land in fertile agricultural regions of other nations.

There may be an element of truth in this argument, but by far one of the most successful means of helping numerous smallholder farmers has been through the application of microfinance. This idea was introduced in Bangladesh by an academic, who left his ivory tower and rolled up his trousers and started to plant rice and work alongside the peasant farmer. He quickly understood their real needs and applied his mind to finding a solution. The result was the founding of the Grameen Bank in Bangladesh, which commenced by providing small loans to smallholders to finance their next crop before the money was repaid. This exemplary organisation has spawned similar activities in forty different countries as its sphere of influence extends round the world.

One of the biggest problems facing those who lend and those who loan is the risk of crop failure when, despite all sorts of helpful advisory and extension work provided by the finance providers, the weather pattern proves unfavourable. Being small scale, poor farmers they have no fat to survive such a blip in fortunes and the loan remains unpaid and the financing of the next crop uncertain. This issue has been addressed by forming farmers into groups or co-operatives whereby there is internal insurance meaning that if one farmer fails to repay a loan then the others cover for the deficit. Other microfinance providers have linked in to larger corporations who require the crops for further processing or trading and have used their financial muscle to provide the insurance. Even further along the insurance chain other microfinance houses have adopted quite complex developed nation insurance techniques whereby the premium is built into the loan and provision of inputs. By protecting the farmer through some form of insurance these growers have gained confidence to invest more into their crops and then to benefit from the delivered improved yield.

We therefore have food producers farming anything from a small patch of poor yielding land to a mega sized corporation growing across mile upon mile of fertile soils. When we are hungry there is not time or worth in arguing about which shop or stall we should buy our food from. The imperative is to provide the science, the market information and the funding to allow each of these farmers to provide sustainable food for the growing population.

Hunger

One in six of us who walk this planet is hungry. Of these the majority are found in the Asia and Pacific region where 642 million are estimated to fall into this category. The other major region where hunger is acute is sub-Saharan Africa with its 265 million hungry souls representing 32 percent of the population. It is perhaps not surprising that the UN considers food security as one of the most critical peace and security issues we currently face.

If we created a single file soup kitchen queue for those who are hungry each day on this earth it would stretch to the moon and back. On top of that there is the real concern over whether world agriculture has the ability to supply that kitchen. The consequences are self-evident and an additional 100,000,000 people are being added to the current level of 1.1 billion hungry persons each year. The US government has reviewed its position and is now focused on improving food production at point of need rather than endlessly shipping huge quantities of surplus home produced grain, calling it relief food, around the world to meet a never ending series of crises. The Intent is to 'reverse the trend of failed agriculture and malnutrition leading to hunger and insecurity'.

This initiative is one of a series of such reactions in part generated by the activities and lobbying of the Bill and Melinda Gates Foundation. The World Bank has launched its own global and food security fund and the G8 committed itself under the L'Aquila Food Security Initiative to finding $20 billion to help build a sustainable base of agricultural production to service the planet's poor. The importance of such a reversal of policy is immeasurable provided of course it is delivered.

In 2010 the cost of world food imports was 11.5 per cent higher driven by the inflated cost of dairy, meat and vegetable oils. This has to be measured against the 20 to 33 percent fall in the price of staples such as wheat, soya, maize and rice. In other words even with large sectors of the primary industry showing significant price reductions the overall cost of food continues its inexorable rise. For us in the West where food costs only represent about 12 percent of our disposable income, this is manageable without too much inconvenience or change in lifestyle, however where there is poverty and most and often all income is consumed by the need to purchase food then price inflation can only be met with reduced intake; a euphemism for hunger.

Sustainability

It is so very easy for us, who live a span of just a few years, to ignore nature and its long term strategy of forming matter over many thousands and millions of years. This disparity of rate describes the difference between sustainability and destruction. The natural and geological world operates on a sustainable model for to do otherwise would mean that it could never have evolved and there would be no existence or 'now'. We, on the other hand, operate under a destruction model; 'you can't spend it when you're gone' mentality. There is perhaps no better example of this than soil. To most of us it is the dirty stuff that we walk into the house or that covers our fingers as we rush to the tap to wash them clean. In fact it is the most precious 'stuff' on the planet that has taken millions of years to form.

The soil gives a very thin cover to the planet and consists of ground down rocks mixed with decaying organic matter. This mix is on average just three feet deep, but of course varies from a few inches to many feet in depth. The fertility of the soil is heavily linked to its organic matter content as old vegetation decays away and bestows the earth with structure and much of its fertility. A well structured soil holds the optimum amount of air and moisture for root development as well as protecting it from wind and water erosion. The whole process of soil generation is therefore continuous and when the degradation of soils exceeds the rate of natural creation of soils, then the planet's food producing potential is diminished.

Human intervention has markedly accelerated soil degradation, as removing natural vegetation to convert land into farmland immediately leaves it naked and exposed for part of each season between harvest and the establishment of the next crop. During this period, often of several months, the soil is extremely vulnerable to the erosive forces of wind and water as well as allowing the ploughed or stirred up organic matter to be oxidised away into the atmosphere.

The most referenced instance of this phenomenon is the American dust bowl, when the grassed prairies of the mid-West were ploughed and cropped, reducing the soil's organic content and removing its deep rooted grass covering, until the terrible sequence of drought and winds between 1930 and 1936 caused it to be blown away in huge dust storms. Over one hundred million acres were affected, causing poverty and migration further west to California. There, initially, the people struggled to make a living as Steinbeck so graphically describes in his book *Grapes of Wrath*.

Whilst the dust bowl episode is well documented because of its sheer magnitude and dire consequences, such events continue to happen, albeit on a more micro scale; even in the UK, black clouds on the fens or sand throws on estuarine lands are common place. When farming in East Kent I benefitted from having very fertile Brickearth soils that long ago blew in from Siberia.

Some archaeologists have argued that loss and exhaustion of soil, due to increasing population and the need to work the land harder and harder, has caused the collapse of civilisations such as the Easter Islanders and the Mayans. This should provide its own warning to the world where exactly the same thing is in danger of happening on a more global scale. North America, with the trauma of their 1930s experience embedded in their collective memory has taken a considerable number of measures to protect its soil, but today it is estimated that there is a loss of 1 percent of this invaluable asset each year; a frightening rate of soil degradation.

Sustainable farming is a matter of working with nature's natural resources, balancing the output or yield from crops or livestock with the inputs required to generate those yields. In essence, the only added ingredient is the sun's energy, which remains renewable for a few more billion years yet and we should treat as eternal.

The well drafted and researched study *Eating the Planet,* commissioned by Compassion in World Farming, compares various scenarios involving differing levels of additional land being brought under cultivation and varying levels of dependency on animals as food providers. The conclusions are unambiguous with a sustainable or organic agriculture being assessed as able to feed the 9 billion plus people in 2050 with the provisos that dependency on animal products is halved, and there is a more even, some would say fairer, level of distribution. The proposed level of nutrition is 3000 kcal per head per day with a protein per diem intake of 80 grams per head; this is a more than adequate for just about anyone.

New ways for old

Farming has managed for 30,000 years to operate using the sun as its sole source of energy input. The growing of food was basically sustainable until half way through the last century when it became oil dependent and started using up the banked sun's energy formed over millions of years and laid down deep in the earth's crust as oil. In a matter of little over 100 years we will have used up the energy accumulated by trillions upon trillions of once-living organisms and released the accumulated and locked-up carbon into the thin atmospheric layer that parcels our planet like a sheet of tissue paper.

Such a release of energy has industrialised much of Western farming and spilled out over other parts of the world. Large tractors require large farms, the ability to control pathogens, pests and weeds with chemicals has encouraged the adoption of mono-cropping, and controlled environments have delivered intensive and industrial scale animal keeping. Low cost energy has accelerated these changes, distanced people from their farming roots and delivered cheap food. In 1900 about 40 per cent of the UK population were engaged in farming; today it is less than 1 per cent. When I joined a farming estate farm in the late 1960s we had twenty two full time farm workers with the fields producing an average wheat yield of 4 tonnes per hectare. Within a period of only twenty years there were just two members of staff and a working farm manager to cover an increased acreage producing yields of wheat of 10 tonnes per hectare. That is the power of oil.

The fact that we farmed largely without oil in just about living memory is somewhat reassuring and of course many small farmers round the globe are anyway familiar with this type of husbandry regime.

A UK study

In the UK we have 18.4 million hectares of farm land covering 77 percent of the total land area. Of this 5.8 million hectares are cropped with 11.1 million hectares down to long-term grass and rough grazing. Sharing these hectares out on a per person basis we have 0.3 of a hectare of cropped land per person, or put it another way the equivalent of 2.5 people for each football pitch area of farm land. With predicted population growth this will fall to about 0.25 hectares per person pushing our football pitch density towards three people. In addition, for each person in the UK there is 0.8 of a farm animal, excluding poultry.

When I was a director of a company farming in Kent in the 1970s we undertook a study of the food we were producing. Taking the total yield of each crop grown we then applied the relevant calorific values to attain a measure of nutritional value produced. From 1648 acres farmed we estimated we generated 5.6 billion calories of food. Using a Wye College Report, which cited 1 million calories as being the annual requirement per person, we concluded that we were feeding between 5000 and 6000 people from our enterprise. That is the equivalent of 8 people per hectare or 11.5 people per football pitch. Based on today's yields that figure would increase to somewhere between 18 to 20 people per football pitch. That is the potential of oil based agriculture in a fertile area with a good growing climate.

Revisiting this study it is interesting to note that even then, just over 60 percent of the cost of inputs was directly oil related. These inputs, including energy, fertilisers, chemicals, seeds, packaging materials and feed stuffs, accounted for 40 percent of the company's costs with labour at 35 per cent and land and buildings accounting for 15 per cent. In other words the direct energy related share of total business costs was 24 percent. Interestingly recent figures based on American agriculture conclude that some 35 years on the energy share of total farm input costs in that country is 16 percent; this of course is not a like for like comparison as nobody could argue that a small part of East Kent is a microcosm of USA agricultural, nevertheless it does underline the point of oil dependency.

Put another way in the USA the production of 1 tonne of maize requires an input of 160 litres of oil whereas in neighbouring Mexico, working on more traditional cultivation method the requirement for oil falls to just 4.8 litres. It would seem that those of us who consider we are 'developed' have much to learn from our 'developing' cousins.

The global farm

Look on the world as a farm. The biggest issue facing the global farm is shortage of water to germinate and support the growing of acceptable yields. But there are other problems. In some 'fields' the temperature and associated effects of climate change have made the hope of getting a crop less sure, whilst in other 'fields' that same change has driven up yields and opened the door to new crop types. In most parts of the farm, the soil is being degraded, with its toll on productivity. The compensatory activity then, is to open up new 'fields' where once forests, in particular, stood.

The cropping plan is largely driven by markets that enable the wealthier regions of the West and now East to take food from far off 'fields' even where those living there go hungry. Wherever possible oil based products are deployed to increase yields and now this is being linked to the purchase of genetically modified seeds to deliver a form of integrated oil based agriculture. Only where access to such inputs is denied is the 'field' farmed organically or sustainably. Finally animals are being moved off 'fields' of permanent pasture and are being fattened and fed using human foodstuffs under intensive indoor regimes.

The quality of the husbandry is very mixed and many 'fields' remain small and hand worked by a largely female workforce with only the simplest of tools. Other 'fields' have been amalgamated into huge unstoppable areas worked by massive oil driven tractors and harvesters. Some of the 'fields' are producing good yields whilst other lie on the margin between cultivation and bareness.

I, like most farmers, have a dislike for anything uneven. I'm not sure why that should be, but an uneven field of wheat is a poor crop, a harvest of a mix of large and small carrots is evidence of a poorly husbanded crop and a pen full of mismatched lambs is scorned. Yet one of the first lessons I learned in agricultural economics was that good land gave a disproportionate yield benefit and that inputs such as fertiliser are much more efficacious on good soil than on poorer land. Therefore, rather than strive for an even yield the lesson was to focus costly inputs on land with the potential to yield the maximum response. Nature is not fair in the sense of being even handed, but it does offer the means of survival or indeed a good living to those who will share their fortune with those in a less fortunate location.

On this basis, it will become increasingly important to make sure that crops that respond best to fertile soils are planted in these 'fields' and crops that have a prodigious need for water are grown in 'fields' with a surfeit of precipitation. Likewise more drought resistant crops will need to be focused on the more arid 'fields' and those that are best able to cope with poorer growing conditions will need to be planted in the less favoured 'fields'. All common sense, which is the language of the farmer, but of course when that farm spreads over many cultures and nationalities then such an integrated cropping plan becomes the more difficult and questions of food security raise their head. Nevertheless without a target we do not have a goal, and without a goal we cannot score.

The importance of supporting the global farm with a full complement of agricultural research has also never been more important. Sadly, in the UK, successive administrations have tended to ignore where food comes from provided the supermarkets buy the cheapest in the world and keep the cost of living down; a task they have fulfilled with surgical genius. As a consequence many of the great agricultural research and teaching institutions, which fed the UK and the world farm with a stream of new innovations and quality graduates, have been subject to their own age of dissolution. Throughout the world this issue has to be addressed as we turn on the tap to find new and sustainable ways to farm in each of our 'fields'.

Of course knowledge which is not dispersed and adopted is of little or no value. The need for us to revitalise extension work and to share best practice has never been greater. Those versed in African agriculture will claim that there is already a backlog of new and improved techniques that would greatly enhance the continent's productivity, provided the indigenous farmer knew about it and had the necessary incentive and courage to adopt it. The same is sure to apply to other 'fields' of our farm.

It is alarming, both in the developed and less developed areas of the world, how there has been a brain drain in terms of practitioners. The West and East have both seen an ageing of the farmer population and in the United Kingdom the average age of a farmer is just short of 60 years, whilst in Japan this figure rises to over 60 years. In the less developed lands the lure of jobs in the city and overseas opportunities have taken many potential rural leaders and able farmers away from the land that their families have farmed for generations. They too are becoming acquainted with an ageing of their agricultural workforce. Once the young have a taste of the challenge before us and the rewards of producing food, hopefully this migration can be reversed. The global farm needs all the care and attention that a well run holding in Brazil, Wyoming, Uganda, Szechuan or Devonshire already receive irrespective of the adopted farming practices.

The current levels of waste of crop in the field, on its journey from field to store, in the store and then on its onward journey to market or processing factory, and then again in the retail link of the chain and the final consumer's home, build into a frightening figure. Curtail this and then in response to a more aware and responsible end consumer, who eats a full and interesting diet, is more moderate in terms of quantity consumed and food selected, the hope of feeding all increases considerably. Meanwhile we must learn from the organic farmers, who by choice or necessity exercise such skills around the globe and the gestation of a sustainable and sufficient farming system may come towards its painful end with the birth of a low carbon, sustainable means of supplying the people of the world with foods from all its 'fields' for year after year and forever.

Finally it would be an omission not to mention the only global food which is wild, and which we still hunt as did our early forefathers before us. However, the fishing industry has been transformed by technology and it now ruthlessly sweeps up ever more fish out of the wide blue oceans; some trawl nets stretch 40 kilometres in width indiscriminately cleansing the water of life. Scientists working in this sector warn that by 2050, just when food demand is at its peak, the seas will be empty of fishable fish. Such a scenario cannot be allowed to occur and whilst extraordinarily difficult to police, governments have a real and urgent role in agreeing a strategy for fishing from our seas and oceans. This, together with fish farming, then has a future and potentially a sustainable contribution to make.

There is no silver bullet or quick fix, for like most things to do with farming, there is much hard graft to be undertaken to bring about change. If we are to move to a fully sustainable process that can service our community of 9 billion, then all the ingenuity of the scientist will have to be intelligently melded with the time honoured ways of the land forged over many centuries. This needs government and farmer to work in harmony with the new consumer, and even then it is likely to be a tight squeeze. For certain we will find that every crumb counts.

LIQUID LIFE

No water no life

In the very early 1990s I attended a food forum hosted by PriceWaterhouse Coopers in London and recall the head of United Biscuits standing up and talking about the shortage of water as being a long term threat to his business. I do not remember anything else from that meeting save that it was the first time I had heard anyone in commerce make such a statement and it was a bit of a shock.

When I came to these writings I had initially thought that the road ahead was blocked by a shortage of food to feed an expanding population. The more I researched, the greater my conviction that the real issue was not food per se, but rather the availability of fresh water to grow it. Indeed it is forecast that by just 2025 the loss in agricultural production due to shortage of water will equate to the agricultural output of America and India combined. These are shocking statistics.

The World Economic Forum estimate that in 15 years' time the global demand for fresh water will exceed its supply by 40%. This is based on demand continuing to grow at its current level of an additional 60,000,000,000 litres each year. Even then 1.1 billion people have no access to safe water.

If we had enough water, then the path to feeding ourselves and morphing into a sustainable lifestyle seemed possible and not without hope, but where there is no water there is no life.

Not a drop to drink

'Water water everywhere but not a drop to drink' has moved from Samuel Taylor Coleridge's sinking boat to lands stretching round the globe. As floods become more frequent so do prolonged periods of drought and by 2005 it was estimated that 44 percent of the world's population lived in areas of severe water stress causing some 1.2 billion people to be living with a critical level of water scarcity. By 2025 this number is expected to increase to 1.8 billion with a total of 4.3 billion living in conditions of water stress. Whilst the sub-Sahara regions of Africa will suffer the most with the Sahara extending by 40 kilometres southward each year, the industrialised world will not be exempted and the USA and Australia are forecast to be amongst those most affected.

North America is a profligate user of water and the consequences are perhaps best exemplified by the Hoover Dam on the Colorado River in Arizona. Built as a part of a financial stimulus package back in 1936, it now only holds back a somewhat depleted reservoir, which has to be supplemented by water piped in from other areas.

It always comes as a surprise when people first learn about the amount of water that is required to grow a crop or produce meat. A kilogram of rice consumes 3,000 litres of water in its making, whilst a kilogram of beef demands 16,000 litres. At the other end of the scale 1 litre of milk requires 1,000 litres of fresh water and so on. It is perhaps not so surprising, based on these figures, that 70 percent of the world's fresh water usage is taken up by farming and that so many people live in areas of severe water scarcity.

Not enough work has been done to fully understand the issues around the supply of fresh water, but evidence of the increasingly prejudicial effect of water shortage is self evident and requires action now if food production is to be maintained and increased. Most water is derived from the sub-aquifer where it has been stored from time immemorial. As we extract water faster than it is being replaced, this supply is reduced and we drop our extraction pipes deeper and deeper until the fresh water, often floating on saline water, is finally exhausted. Once again humankind is living off the past, but it is an heirloom that is rapidly dwindling. This is a global problem and even in the UK with its relatively wet climate, the issue of sub-aquifer exhaustion is very real.

Mining water

Irrigated land produces approximately 40 percent of the world's food from just 17 percent of its land. In other words irrigated lands are several times more productive than land solely dependent on rain. Now look at this on a world map and it can be seen that the 2050 world' dependency on food will fall heavily on the Northern Hemisphere's temperate areas, with equatorial regions and the Southern Hemisphere being exposed to water shortage and thereby nutrition shortfalls, as a result of drought and reduced quantities of available water.

The water locked up in snow and glaciers accounts for 10 percent of the world production of crops. This long term and winter storage of water, which is released during the melt of the warmer growing season, is now under threat from climate change and the loss of solid water mass is rapid. Researchers suggest that the adverse affect on agricultural production could be catastrophic to dependent regions. Initially through the period of rapid melt there will of course be ample availability of water, but quite suddenly, that supply will decline until torrents turn into trickles. The Himalayan glacial melt water alone feeds land that produces 25 percent of the world's wheat production.

Buying water

The shortage of fresh water has already impinged on food production and has focussed the minds of academia and governments alike. As food supply tightens, as evidenced during the 2006/7 crisis, countries are becoming increasingly concerned about the feeding of their populations. Any shortage of food immediately triggers unrest, leading to riots and political instability and all governments know they have a prime duty and a vested interest in ensuring that those whom they govern or represent are adequately fed.

For some countries like Saudi Arabia with little fertile land, and more critically no water, the situation has become acute. They have already announced that by 2016 all wheat growing will be halted in order to preserve sub-terrain water supplies. As a consequence they have adopted a policy of investing in land off-shore, and have established these farming bases in Africa and East Asia, initially to grow rice and cereals before expanding the repertoire to cover all the food staples. The Saudi contribution is to supply the seed, farm machinery, storage facilities and supporting infrastructure and in some circumstances support the building of schools and medical facilities. However there are political consequences and when a large tract of Pakistan was recently acquired by Saudi Arabia with a concomitant investment in infrastructure supported by the services of a private security force of 100,000 to protect that investment, it was classed as the 'new colonialism' by activists with perhaps some justification.

China, with its burgeoning population, has also engaged in off-shore farming buying up tracts of farmland in Africa in order to secure food production and supply. Along with their other investments in minerals, this now makes them the largest sovereign investor in this economically emerging continent. It is estimated that there are over one million Chinese people living in Africa; a large number of them are engaged in farming these newly acquired lands. But China and Saudi Arabia are not alone and the number of countries involved in this activity has grown and now includes India, South Korea, Egypt, Libya and the Gulf States. The number of these deals, whereby farmland is bought or leased, has escalated to the point where the FAO reckon that 20 million hectares are farmed off-shore or to put that into context the equivalent of the farmland of two Germanys. As Mark Twain famously said, 'Buy land, they're not making it anymore'; as a rider they certainly don't make land with water. The result it is that the value of this scarce resource can only rise and will increasingly be sought after.

These land grab investments will both threaten the indigenous farmers, who often have no title to the land they have farmed for generations, and bring benefits to the communities they affect. The question is whether the benefits will outweigh any adverse effects and irrespective of that balance whether the system is stable in periods of food shortage or political unrest; in either case this could lead to conflict. Arguably, on the positive side, the example of Egypt's acquisition of two million acres of land in Uganda for the production of wheat, may go a little way to allay the tensions over water between these two countries as they share the course and water of the Nile.

In essence growing off-shore is a means of accessing fresh water. It is a simple case of taking the mountain to Mohammed and rather than using in-country water to grow crops, opting for using other people's water to produce food. Professor Jon Anthony Allan has expounded the theory of virtual water to elucidate this strategy. Virtual water is the volume of fresh water required to grow a specific crop; it is virtual because in a tonne of say wheat the actual water content weighs just 140 kilos, but the 1300 cubic metres required to grow the crop is no more and is therefore virtual.

An Indian solution

India is one of the countries at the forefront of the water crisis, as their water table is falling at an alarming rate and up above, at soil level, crop production fails to feed a rapidly growing number of people, which now represent 14 percent of the global population. Nearly 84 percent of India's water is used in agriculture, irrigating half its 140 million hectares of farmed land. The other half is totally dependent on rainfall, with any shortfall in precipitation resulting in crop failure or yield reduction. The traditional methods of irrigation, using canals and wells to inundate or often manually sprinkle crops is inevitably pretty inefficient, as poor targeting and high evaporation rates in the high temperatures reduces efficacy. Indians have been farming this way for centuries, if not millennia, and yet sitting at their door was a far better watering technique now widely used, but first developed by an urbanised people forced to farm in order to survive.

When Israel was established in 1945 after the Second World War, many Jews from Europe's damaged cities left years of persecution and worse to settle in this new land. These were primarily town dwellers more familiar with law, finance or trade rather than farming. When visiting Israel I was told the story of how these new farmers, faced with feeding their compatriots on a Kibbutz, were sent a goodwill consignment of vines from France. Unsure how to treat this apparently dead looking root stock they carefully followed planting instructions getting the row widths and plant spacing just right. However, the plants did not grow and eventually they sent for an agronomist, who was somewhat surprised to find each plant, whilst perfectly spaced, had been planted upside down with the roots in the air. That was their level of know-how when they started out and yet very quickly they looked at their dry and often stony land with fresh eyes and equally quickly learned that crops need regular watering if they were to thrive. Water is something Israel does not possess; the river of Jordan is not 'deep and wide', but just a trickle.

The result was the ingenious invention of drip irrigation which meters out regular but small volumes of water at the base of each plant's stem via runs of narrow bore flexible pipes. Recently India has embraced this technology, no doubt with their own effortless way of embedding economy into its manufacture and operation, so that already an estimated two million hectares are irrigated using this technique. The saving in water is a good 50 percent when compared with the more traditional means of irrigation with the added benefit that yields have gone up several fold in a quite dramatic way; a real win win.

Liquid life

In the business world, where there is a scare resource or one that is geographically unevenly available, then the situation is appraised and enquiry made as to how the return from that scarcity can be maximised and how the product from productive regions can best be distributed. Different crops have different levels of demand for water and the logic is to grow those with a moderate rather than excessive requirement. Similarly there must be logic in growing high water demanding crops in areas where there is ample natural precipitation or water source. Once again, as a recurring theme, the logic of growing meat in areas of low water availability must be seriously questioned as of all food stuffs this is by far the most demanding on fresh water. Nature has known for a long while that grazing animals require grass and grass grows where there is rain.

The need to make this type of management decision can then be supported by a whole array of technical support mechanisms. Some of these techniques and systems will deal with how water can efficiently and most effectively be delivered to the plant's root system. Other techniques will address the plant's requirement for water and how this can be curtailed through breeding programmes. In addition there will be a continuation of the time honoured search for ways and means of holding stocks of water in dams and reservoirs that do not prejudice the local ecosystem or cause offence to the downstream neighbour.

However we look at the problems ahead the one recurring theme is the availability of fresh water. There is not as much around as we have perhaps always assumed. Climate change is radically altering its distribution and its level of natural storage as ice, and agriculture is consuming greater volumes, driven by increased demand for crops and the production focus on products that are inefficient in their use of water and with high demands for this liquid. We must safeguard fresh water with our lives for without it there is no existence; water is nothing more or less than liquid life.

TRANSFORMATION

Reciting the past

Looking back is both educational and essential if we are to address the future, for it is the root system that dictates the size of the plant, and our root system is buried in our past. However, looking forward is hazardous because as the financial adviser always solemnly adds 'past performance is not necessarily a guide to future performance'. As we are dealing with the really serious issue of our survival as a global community, we must weigh the balance very carefully to ensure that the framework in which we are to operate is at least sound, even if the inserted detail proves almost inevitably less so.

Going back three and a half decades and sitting in the presence of E. F. Schumacher, a guru and visionary of his time, I listened to his eloquent paper dealing with the UK and the future issues it was likely to have to face. He painted a picture with eight predictions and twelve proposed ways forward, some of which were both engaging and, at that time, quite revolutionary and challenging.

Predictions:

1. North Sea oil supply would be heavily restricted by 2004 as the UK hit its own peak oil point.

2. Coal supplies would be difficult to increase over and above 1970s levels.

3. *Atomic power is expensive and potentially dangerous and therefore little progress would be made on expanding this source of energy. The energy required to build an atomic power station absorbed the equivalent of 15 years of its energy output. With a life of 25 years this left a net energy benefit of 10 years worth.*

4. *Population will increase well into the 21st Century.*

5. *Availability of imported food will fall away as world population doubles by 2000 and then continues to grow.*

6. *Nitrogen fertilisers, which are energy hungry, will experience significant price increases linked to the increased price of oil. Phosphate fertilisers will become in short supply in the foreseeable future.*

7. *The importation of protein for animal feed will be less viable as the world protein shortage increases.*

8. *Supplies of fish from the seas will become restricted with over-fishing.*

Way forward:

1. *Britain needs to increase its food self-sufficiency.*

2. *Agriculture needs to become more energy efficient.*

3. There is a need to start re-cycling of all organic waste as a source of fertilisers.

4. The consumption of meat needs to be reduced.

5. Meat production needs to be concentrated on the hills with reduced focus on intensive lowland systems.

6. Factory farming is wrong as it creates health risks to animals and humans, pollutes and diverts food for human consumption to animals.

7. Livestock farming should be re-integrated with arable farming to maintain fertility and reduce energy utilisation.

8. There should be less dependency on chemical herbicides and insecticides through the use of integrated crop management and biological control.

9. Refocus on smaller farms that have a higher output per acre.

10. Encourage the cultivation of allotments.

11. Develop organic farms linked to small scale food processing units that encourage people to return to the land.

12. It is essential that we adopt a spiritual basis for our lifestyle and a way of production that is in accord with real human needs and the will of God.

His predictions have largely been proved accurate and his means of dealing with them still lie on the table.

The debate that followed his spiel was energised as the forces of industrialised farming met the more cerebral stance of Schumacher's more visionary school. Those of this latter tendency argued that we would become less wealthy as a nation and would therefore be unable to sustain the continuing importation of food. They believed that unemployment would increase and that this would create an opportunity to employ 15 percent of the population on the land making a more balanced and healthy society. This in turn would need to be supported by a career structure, or farming ladder, with smallholdings providing a starting point for progression to larger holdings albeit small in today's terms. Agriculture, they also concluded, needed to get off the industrial band wagon and meet the needs of the people. Those more inclined to Schumacher's position, argued that industrialising agriculture was gearing the industry to deal with inanimate situations, whereas agriculture is concerned with life and the uncertainties of living ecosystems and of the climate. This point was driven home with the warning that such industrialisation was doing violence to those ecosystems, which may prove to be harmful to them and to mankind.

The conventional or industrial, large scale farming lobby hit back with slightly muted comments concerning the risk to food output arguing that technologically driven husbandry increased yields and quality of crops, that the lack of availability of trained labour already threatened production, and that their ways were the best means of producing cheap food for the masses.

The two camps seemed a long way apart and indeed the debate on sustainability still continues to rage, with those supporting the status quo of oil based farming maintaining this is the only route to feeding the world, whilst those arguing for sustainable farming, hold that their way forward is the only means of being able to continuously feed the world way into the future. The protagonists shared the same initial goal, but the proposed means of its achievement were by very different strategies.

Sustainability versus growth

Sustainability and growth are not good bedfellows. In many ways they are almost contradictory. Part of the problem is that man, and particularly the gender man, has acquisitiveness and growth built into his genes as he cannot help himself when it comes to building bigger barns or forever wanting to go faster; boundaries are there to be broken.

Once, a salesman selling a piece of extremely expensive and sophisticated food processing equipment, recounted how he had installed a similarly complex piece of kit into a factory making custard cream biscuits. Knowing that the machine was well proven, he was somewhat surprised that despite having taken considerable care to ensure the operatives ware fully trained in its use, the new machine kept failing. When his engineers attended the site there were no problems and yet as soon as they left the phone would ring again with yet another breakdown to report. Further investigation brought them to the conclusion that the machine minders were simply getting bored and then twiddling the knobs to increase the speed of throughput. Machines don't think like that and just want to work day in and day out undisturbed at the same speed and in the same rhythm. The maker's clever response was to fit a new dummy control box on the machine that allowed the minder to believe he was making all sorts of alterations, without in fact affecting the machine. The result was satisfied operatives and a machine that stopped breaking down as it quietly and secretly continued its pre-set and fixed routine.

But whatever our inclination, the science and logic is that we have no choice but to embrace sustainability. The planet is a store and however bounteous and forgiving it has proved in the past, we have now reached a point where we are consuming what it holds more quickly than the sun, with all its energy, can replenish it. That is an amazing feat of consumption, but one that is already giving rise to 'aposematic' events and which can only lead to an apocalyptic conclusion. We therefore have no choice but to change our consumption pattern and bring it back in line with nature's ability to replenish; in so doing we move to a sustainable way of living.

Transformation

If we can argue, albeit with provisos, that the earth can sustainably feed and service a population of 9 billion at least in the short term until that population declines to a more securely sustainable level that has a greater margin of safety, then we need to manage this transformation. But this is not a spreadsheet exercise and it will require change in human activity, preferences and tolerance to that change. So whilst the availability of fresh water may limit food production to the extent that it has to be managed and rationed in order to fill every belly every day, that management will require buy-in by governments, corporations and the individual. If that is beyond us then so is universal survival.

A flawed example on our doorstep

To take just one example to demonstrate how we need to adjust and to bring the argument home to our own backdoors, let us consider our means of acquiring food in the UK. This example is not a comfortable one for me as I have a huge respect for many friends and colleagues in this sector, who work extremely hard under considerable pressure to bring the consumer a better shopping experience and excellence in service. Here the focus is on the model they have adopted rather than the organisations themselves. In terms of the environment that model is heavily flawed and requires a radical and constructive rethink.

When I was in my late teens I first went into a supermarket. Set in Bromley High Street it was called *Carters* and was an advance on the *Pay and Take* store that we had used regularly. The experience of picking goods off the shelf into your own basket before paying at a checkout point was exciting and novel. From then on the rise of the supermarket has been unstoppable with the contemporaneous explosion of out-of-town shopping and car ownership causing a revolution in our buying habits and lifestyles.

Initially these new outlets sold the established and familiar brands of ambient grocery items alongside their fresh produce offering, which was locally sourced. The ability of the supermarket to use its size to extend its offering, buy competitively and sell cheaply was of course massively appealing to a generation who still remembered the privation of the 1940s and 50s. It was not long until these supermarket chains started to offer their own label goods, which undercut the brands and so began the transfer of consumer trust from primary manufacturer to retailer. The offering widened with the growth in affluence and the ability of the stores to provide chilled cabinets for perishable foods. The use of local suppliers was abandoned as quality of produce between stores varied and the system was open to the passing of backhanders. Instead, all goods went through central depots, or RDCs, where they were counted and checked in before being counted out and dispatched to store.

A person's lifestyle was defined by the store they used. Tesco's founder, Jack Cohen, famously chose to 'pile it high and sell it cheap', Ken Morrison, son of the chain's founder, retained tight control by locating all his stores in the Bradford area with none more than half an hour away from his office. M&S cherry picked premium items for top up shopping focusing on quality and product innovation as its brand message, and J Sainsbury transformed itself from a grocer to the middle class into its supermarket format, whilst shaking off family control.

The whole sector therefore pandered to the British class system making us feel at home. The upper end of the middle class were M&S and Waitrose shoppers, the main stream middle class were defined as J Sainsbury and Safeway shoppers, whilst the lower income shopper used Tesco, Morrison and Asda. Each chain had its own suppliers list and loyalty to them provided a degree of security and a rationale for investment right down the supply chain. Meanwhile shoppers responded by demonstrating fierce loyalty to their chosen supermarket and all seemed right with the world.

The entry of American based Wal-mart into the UK, when it took over Asda in 1999, fundamentally changed all this; slowly at first and then at an increasing pace. Their aggressive penetration into the mass-market sector challenged the other supermarkets, with Tesco responding very effectively, whilst Safeway disappeared off the map. Sainsbury initially lost market share as did Morrison as they stumbled trying to digest their acquisition of Safeway. Add to the pot an evolving oversupply of supermarket outlets, the almost complete demise of the independent sector, the rise of internet shopping and an increasingly streetwise shopper, who had been transformed into a consumer and the result is unbridled competition that has now moved from being chronic to acute.

That is the background that has formed or allowed consumerism to develop in the UK. A 'shopper' used to buy for personal and family needs according to their household budget and wage packet. The 'consumer' on the other hand is a professional buyer, who can access funds via credit cards, who comparatively shops, and who can fulfil any demand for goods or services immediately. Consumers buy their lifestyle whereas the long gone shopper's only intent was to maintain existence with a few occasional treats thrown in.

Having worked with the whole gamut of supermarkets I am conscious that they have brought new freedoms to the market place. The opportunities for their customers to access new and innovative products, convenience products, products from around the globe, quality fresh and chilled products and one stop shopping has brought a new perspective to the shopping experience. Their demand that, at minimum, their food meets regulatory standards in terms of nutrition, hygiene and purity has been uncompromising. To these attributes they have now added ethical standards relating to suppliers' management of labour and their procurement of product. This means that exploitation of workers in the UK and beyond to the least developed and ill-regulated of countries has been minimised, certain species of fish or animal under threat of extinction have been withdrawn from shelves, and relatively high animal welfare standards have been adopted. Of course they are not perfect in this regard, but I was impressed how companies like M&S has come to terms with campaigning groups and over the past decade has positively collaborated with them to meet their demands and in so doing, improve their own ethical position and thereby their business security for the long haul.

Whether the national multiples can be sustainable or improve the levels of sustainability is, however, a very different question. The problem they have is borne out of their size; and with supermarkets market share and size is everything. Their business model has to be low margin coupled with high turnover in order to protect their competitiveness and to build their customer base. Tesco has pursued this business model extremely effectively and now £1 in every £8 spent in the UK rattles through their tills.

But to be so successful and to perfect their model, the supermarkets have had to centralise their distribution so that goods travel many internal miles. They have trawled the world for cheaper food sources adding many international food miles to products. They have located their stores for ease of access by adding car parks full of domestic mileage, and they have had to package their foods to assist shopper handling, improve display and attractiveness, and speeding the check out process. All this has militated against local suppliers, the minimisation of the carbon footprint and the minimisation of waste material disposal.

As a supplier of fresh fruit products to the supermarkets we would draw in the fruit from the overseas growing regions by sea or road and squeeze and bottle the product or prepare and cut up the fruit in the UK. When freshly squeezed and retail-ready pineapple juice was introduced from a Ghana based supplier, we were prompted into looking at our own business again. As a part of that review we undertook a study to assess the comparative environmental impact of flying in finished product from overseas as against shipping in fruit by sea or road and processing it in the UK. I was shocked when the internal paper was presented. Bringing 1 kilo of strawberries from Spain by lorry required 0.17 litres of fuel oil, however flying them in from Israel required 1.9 litres or nearly twice the weight of strawberries in aviation fuel. Even worse was flying prepared fruit from South Africa, which required eight times more energy than processing the same item in a factory in the UK. Finally, the Ghanaian pineapple juice required 1.37 litres of fuel to fly in just 0.5 litres of juice, when to produce it in the UK required just 0.16 litres.

This is of course madness as moving produce around by air is forty times more harmful to the environment than using sea transport. There are well rehearsed arguments that flying in labour intensive perishable foods from warmer climes, often in developing countries, is good for the local economy, but the global warming impact of this behaviour will only condemn these same countries and workers to intolerable hardship. Far better to rethink the model and seek out a more environmentally sound business for these important and enthusiastic teams of staff.

Perhaps the most graphic summary of what is really going on was given when the food miles travelled by a trolley of supermarket food stuffs was analysed; when added together it was equivalent to a one way journey to the moon. It is doubtful whether the supermarkets, with the best will and intent in the world, can change their environmentally damaging impact; it is scored into their raison d'être and business model.

To further evidence this rejection of the supermarket model, there is the issue of the move to fortnightly removal of household waste. This has proved very controversial, partly perhaps because it has highlighted the amount of waste we manage to generate. Working on a two bin system, one for recyclable materials and the other for landfill items, I was alarmed that every alternate week the recyclable bin left our house full. We made a New Year resolution to reduce this frequency of emptying to once every four weeks, or in other words to halve our waste accumulation.

We tried to reduce the amount of packaging we were collecting up at the local supermarket, but despite our best efforts this proved difficult and not particularly effective. The chaos caused by loose produce, freed from any plastic bag and rolling around the checkout belt, defeated the system and caused problems for the assistant and those in the queue behind us. As a consequence we made a concerted effort to stop shopping in the supermarket and principally bought from the local shop, the nearby farm shop and the weekly street market. The difference in the amount of packaging waste generated was startling and the objective of halving waste easily met.

This simple exercise provided its own irrefutable evidence that the supermarkets have driven waste volumes. It is true that they have worked hard at converting much of this to recyclable material, but it still requires energy to manufacture it, then to transport it and finally carry it away to the processing plant where more energy is used to complete the cycle.

The supermarket model has other shortcomings and the length of their supply chain is just one of these. In the 1970s, working with two colleagues, we introduced Iceberg Lettuce to the UK market. Up and until then the wilt prone Butterhead lettuce and short season Cos lettuce were the main constituents of the salad bowl along with cucumber and tomato. Perhaps it was not surprising that salad consumption was pretty low and the only real hope of eating a good salad was one cut and made from the garden or allotment and immediately consumed.

The iceberg lettuce had been developed in America so that lettuce encased in crushed ice could be shipped many hundreds of miles from the growing areas of California. At the time, the only iceberg lettuce available in the UK was flown in twice a week from the West Coast of America and was of course extremely expensive and confined to the upmarket hotel and restaurant trade in the capital.

Transferring the Californian growing and harvesting techniques to this country was not easy for we worked under a different climate regime and at a substantially different latitude that affected day length. After intensive trials of fifty two different varieties, much hard work and many failures we commenced the production of iceberg lettuce in earnest. The consumer reaction was explosive and there seemed no end to the demand for the product, as each day we harvested the crops using mobile rigs sweeping across the fields and on which the heads of lettuce were wrapped in film and boxed, according to size, before being whisked off to the central vacuum coolers in old army four-wheel drive lorries, driven by often overzealous and under-experienced students. The vacuum cooling process worked on Boyles Law whereby the palletised boxed lettuce were trucked into a large cylindrical container, the heavy door was closed and sealed before the air in the container was extracted. As the vacuum was pulled so the boiling point was lowered and the cooling occurred as the moisture on the lettuce leaves boiled off; the same effect as tipping a little ether or solvent on your arm. The lettuce at or below 5°C, was then sent by refrigerated lorry to the supermarket depots around the country, from Kent to Scotland and over to Ireland and Germany.

The supermarket reaction was predictable and for them to have iceberg lettuce on the shelf during the UK season from late May to October was not enough; they and their customers wanted it year round. Our response was to research Europe and to make the decision to grow the crop in Spain through our winter. A Spanish company already grew iceberg in the Murcia area and such was the value of their market position that they had rather large surly men on motorbikes patrolling the roads running alongside their fields. Consequently, on the several occasions we stopped the car to try and get a better view of the crops one of these motorbikes would come alongside before we could even get out of the vehicle and make it very clear that we had no option but to move on.

In the end we found a very good Spanish business partner and established a growing base near Murcia and each autumn we would ship down harvesters and vacuum coolers to maintain supply until the following May when they would be shipped back, serviced and turned round for the UK season. That was my first taste of maintaining year round production and it was a model that I was to follow, when a few years later we launched the bagged washed lettuce and sourced the various types of leafy salad produce out of France, Italy and Spain during the winter and spring months, when UK production was unavailable.

With insider knowledge of the industry and knowing the lengths the supermarkets will go to maintain product flow, it is not surprising to me that food for the average UK household is the main contributor to their carbon footprint. This was further exacerbated by the introduction of Sell By, Display Until and Best Before dates that started to appear on both short and long life goods. This practice makes eminent sense where food is chilled and has a short life before potentially becoming susceptible to the harbouring of pathogenic organisms. When, however, it is applied to other food stuffs, the logic is less compelling and repeated studies have shown that it leads to considerable quantities of good foods being disposed of simply because of an often arbitrary date on the packet, jar or can.

For the weekly supermarket customer the carbon demand of the supermarket system is most probably a little surprising and anyway by its very nature of being a staple of life, it is easily excused as being essential and quickly dismissed from the mind. The supermarkets make great play of their green credentials, which further calms the conscience of even the more concerned shopper. Some companies have certainly been responsible in terms of using only sustainable sources of fish and wood and the Co-op in particular has worked hard at extending their organic offering. However, none of these multiple outlet businesses has really addressed the core issue and for good reason. The problem is that the business model they use is environmentally flawed and incapable of being green and despite their best efforts to ameliorate their position, they can only make the best of a bad job. Of course this is much better and more responsible than not caring, but regrettably does not provide a sustainable answer.

Perhaps one of the more easily observed indictments of the supermarket is that when they do move into an area they kill off the incumbent independent traders. The argument flows that it is the consumer or the inhabitants of that town who change their shopping habits and loyalties and transfer them to the supermarket. This is partially true in so much that independent retailing is a knife edge business that only has to lose a relatively small amount of business for it to fail. Of course in any community there are those who are hard pressed financially and who have little choice other than to seek out the lowest priced goods, or those who are time pressured and prefer one stop shopping based upon extended opening hours. It, therefore, is inevitable that as these people migrate to the new supermarket the survival balances all along the High Street creak in the wrong direction and shops close and businesses fail. But we should not forget that for every pound sterling spent locally 37p is returned into the community, whereas the same amount spent with a national organisation will only return 12p.

The most difficult challenge

The supermarket issue is just one of many equally important questions regarding our love affair with excessive travel, our extravagant lifestyles including over-eating, over-heating/cooling and over-consuming and so on, which all lie at our door or in the pending tray. The question each of us faces is – are we willing to change; change our shopping habits, eating habits, heating habits, travel habits? That in a nutshell is the key to change, a very simple question and the most difficult to address.

It is not so difficult to have a change of mind-set once we are shocked into action and respond emotionally, or alternatively are enlightened by argument and respond intellectually. The former is imposed upon us and is outside our control, whilst the latter is a matter of our choosing and is therefore very much within our individual sphere of influence. Of course the shock or the conviction has to hold enough gravity so that it can stand up to the survival bent energies of business and interested parties. They will inevitably try and persuade, cajole and bribe us, the consumer and citizen, not to revert to being shoppers rather than consumers and will use their honed skills to persuade us from becoming more society-conscious and aware people to the detriment of their business.

That transformation will have myriads of means of transforming our lifestyles into more sustainable ways. Some will be small, some of greater significance, but all will be of critical importance. To use E F Schumacher's words 'to talk about the future is only useful if it leads to action now'.

This book, to parry Isaac Newton's words, is an extremely small 'pebble' on a long beach full of pebbles and running down to the huge ocean before us. If nothing else, it has tried to prise open a little further the question before all of mankind – how are we going to feed 9 billion people sustainably? It is a massive subject, webbed by complication that makes it largely impenetrable to the human mind. Despite this there are some very clear pointers in the argument and thought process that do not require the sophistication of the academic's mind, but rather the application of common sense. It is these common sense matters that should provide us with our initial focus. In the days of Clinton's election to the Presidency the scrawl on the wall was 'it's the economy stupid!' and with Obama it was 'change we can believe in'. Now we need our own scrawl on the wall that will propel us through the next difficult phase of our history; a scrawl that speaks as much to the individual as it does to the global community of nations.

What about 'I can change the world'.

Finding common ground

What then are the big conclusions that we can draw from our trodden path? Firstly we need to define the common ground so that we can share the land, agree the start point and set up our base camp.

The Two Principles

1. The world's human population is increasing almost exponentially and provided such growth is not interrupted by a *force majeure* it will reach 9 billion or thereabouts by 2050.

2. The earth's atmosphere is currently going through a phase of temperature increase.

The Consequences

The consequences of these two factors of change are partly known albeit in the greater part unresearched, but again in order to maintain our focus we need to identify those prime consequences that are having or will have the greatest effects.

1. Extra people put further pressure on the human habitat. This raises concerns about having enough to eat and drink, having adequate living conditions, the opportunity to be gainfully employed and the means of providing for the sick and elderly.

2. Human migration will increase as the search for food and water intensifies and as a consequence peoples move to more temperate regions.

3. Temperature increase is responsible for driving concomitant and fundamental changes to the climate.

4. Climate change is in turn having a major effect on the distribution of fresh water availability both in terms of its pattern, natural storage in lakes and as snow and ice, and as localised quantities of precipitation resulting in greater frequency of deluge or deficiency in any one location.

5. The combined consequences of temperature increase and changes in fresh water availability are having a profound effect on the availability of land for food production. New acreage is being found from deforestation and in areas that are becoming more suited to crop growing as temperature increases. Fertile land suitable for cultivation is being lost as a result of desertification and soil loss, sea level rise resulting from ice melt, and temperatures rising above those able to support vegetation growth.

6. Growth in global food production, having enjoyed a period of rapid increase, is now levelling off as climate change consequences come into play. This slow down, exacerbated by population growth, has increased the number of people who go hungry; an increase of 100,000,000 each year.

7. Climate change and the incursion of humans into many different habitats have created ecosystem pressures, resulting in loss of distinctive life supporting habitats and consequently a rapid increase in the rate of species extinction. This is currently estimated as being a thousand times faster than the normal level of species extinction.

Causes

The reasons for all these major occurrences then becomes important to our process in that it enables us to address these causes as part of our damage limitation exercise, before we move on and allow nature's own reparations process. Once we place these on the table the debate stirs for they bring with them the baggage of scientific proof, questions of morality and partisan interests. Again to avoid walking into a wall of controversy where momentum is lost and energies dissipated resulting in a loss of focus, it is essential to the process to seek out as much common ground as is possible.

1. The reasons for population growth are a mix of increasing population of fertile women resulting in a higher birth rate, and lengthening life expectancy.

2. The increase in atmospheric temperature correlates with the increasing levels of greenhouse gases in the atmosphere. These greenhouse gases are linked in part to the activities of man.

3. Deforestation in response to commodity demand and the search for new land to farm, is responsible for a fifth of greenhouse gases as the size of this carbon depository is reduced.

4. The shortfall in food supply is linked to having more mouths to feed and a slowing in the rate at which increases in food production can be achieved.

5. This shortfall in food supply is further exacerbated by changes in the global diet as increasing quantities of meat and animal products are consumed.

6. The current and projected growth in animal farming will test the ability of the planet to support all of human life sustainably. Animal production is demanding on water use as well as land use and generates a disproportionate volume of greenhouse gases.

7. The combination of population growth and limitations on growing areas, has already resulted in, and will continue to result in, soil degradation and loss. This is, in the human timescale, an

irreplaceable asset and consequently the fertility of the earth will decline.

Action

We then come to the nub of the issue; what actions should we take to mitigate the effects of the two big changes we have identified that now threaten much of what we know and enjoy? Here again, we need to look at the bigger issues that governments, the food producer and the individual in particular can endorse and exercise, and in so doing, accelerate the changes we need to make.

These have been restricted to five in order that each category can maintain a high degree of focus. This in no way suggests that other mitigating actions are necessarily of lesser value, indeed any activity that helps resolve or ease our current predicament is of inestimable value, for we are talking about balance and to tip a balance takes but a single grain.

Government

1. Health policy should focus on provision of contraception aids and fertility management advice in order to reduce population growth. This policy requires to be underpinned by education and recognition of equality.

2. Governments should invest in the preservation of the forest regions to safeguard this carbon sink and area of rich in species habitat.

3. Fish stocks should be protected from over-fishing through international governmental treaty.

4. Governments should free up trade and remove the distortion of subsidised agricultural to encourage food production in the most agronomically suited areas for each crop type and as close to point of need as is possible.

5. Governments should agree a global level of fossil and biomass fuel tax correlating to carbon emissions of each fuel type. This would be a dual tax to be firstly levied on energy generation and then secondly charged at point of end-product consumption. These tax incomes should then be linked to foster fledgling support for early stage renewable energy schemes and technologies as we move to total reliance on energy from outer space.

Agriculture

1. The industry should reenergise agricultural research with supporting extension activity into productive sustainable agriculture that is sun based.

2. Farmers should be encouraged to produce crops that best match the available water resource, soil type, climatic conditions and local need. Removal or evening out of subsidies assists this.

3. The industry should extend the use of water economy systems to irrigate crops and breed

varieties that are characterised by a lower demand for water.

4. The soil should be protected from erosion, exhaustion and physical damage, through improved husbandry and management techniques.

5. Ruminant animal production should be confined to natural grassland areas and intensive industrial and semi-industrial rearing of animals should be phased out.

The Individual

1. We should reduce and phase out our consumption of 'industrialised' meat.

2. We should reduce food and packaging waste by shopping for need, buying locally, and growing some food wherever possible.

3. We should reappraise our personal lifestyle by cutting out over-shopping, over-eating and over-heating (or cooling).

4. We should minimise fuel usage by local living to reduce food miles, personal routine mileage, and air travel.

5. We should use renewable electricity either via a national energy provider or through local generation.

The outcome

Every generation has tended to consider they are at the end of history or the end of knowledge for the owl of Minerva indeed flies for each passing life or generation and in so doing bestows, for a moment, best knowledge and greatest wisdom. When my youngest son was just four he made an observation about a very wise and knowledgeable family friend who was then well into his eighties, but who retained the agility of mind of a younger man. His pensive observation was simple, 'if there were ninety Johns we would know everything'. There is a real element of truth in this, for if we step aside and again look it soon becomes very apparent that for each generation there is indeed knowledge gain, but sadly some knowledge loss. Nevertheless it is but a part of a chain that stretches from our very beginnings to our eventual end. It is a chain in which we must play a full part and in which we must not become the weak link that causes it to fail and break.

My first job after graduating was as an Assistant Farm Manager. I enjoyed nearly every minute of it. One of the jobs I did was driving casual staff, mainly women, in and out of work. The stories I overheard were educational. One of them went like this, and was told by a rather flustered lady called Anne who spilled it out as soon as she got into the back of my long wheel-based Landrover. The problem was, she explained, that they (meaning her family including three school kids) had agreed to look after the neighbour's budgerigar for two weeks whilst they were on holiday. The children had been clearly excited by this prospect so it seemed a good idea. You can therefore imagine her horror when she had come down that morning to find the bird dead at the bottom of the cage after only a few days under their care. As the perplexed and crestfallen family sat round the breakfast table the terrible truth began to dawn, for the question being asked and re-asked was 'did you feed it?' The response crisscrossing the table being 'no, I thought you did!' So it was that the much-adored bird had died of hunger – not as a result of disinterest or neglect, but due to a simple lack of any one person taking on the responsibility to feed it. That little event speaks volumes as a modern day parable that we need to learn by as we come towards the end of this particular short journey.

So we return to our transformation process hopefully more determined. The proposed key actions set out for government, agriculture and individual inevitably work in some form of consort and share a common theme as they all address the same question; how are we to feed ourselves? That of course remains the goal, but in asking our question the route to its achievement seems also to have emerged. The opportunity we have to feed ourselves and survive is dependent on a series of choices that we as individuals, the markets we make, and the government we elect or tolerate, must make.

What is of real interest is that in all things related to achieving a successful outcome there is time and time again the rationale of the strong and natural imperative of sustainability. To help reduce carbon emissions we must protect the natural forest carbon sink and therefore need to maintain our forest cover by practicing sustainable forestry. To avoid soil degradation and dependence on oil we need to farm sustainably. To maintain fish populations we need to practice sustainable fishing. The list goes on and perhaps not surprisingly, we find that living as we do in a natural world teeming with species, each of which is dependent on sustainability, so we too have to learn, or perhaps relearn, how to practice a life of sustainability from this world; indeed the answer is all around us, right before our eyes. Once we can achieve such harmony, then all of us have the chance of living lives with food, shelter and worth.

Of course there are serious and genuine doubts as to whether the earth can sustain 9 billion people as a long term proposition. Nevertheless, to achieve that or survive the population bulge period within the given time frame will require fossil fuels to see us through the transition, research to take us to new ways of providing sustainable energy and food, the best efforts of conventional and sustainable farming methods to keep us fed whilst we find our longer term sustainable existence, and an acute and zealous protection of habitat.

The post-Modern age of man with its Information Age and Consumerism must now move on to a period of Sustainability or wealth without growth. The Age of Wealth without Growth is a real, urgent and essential goal. To achieve this over the next few decades is not going to be a matter that is perfect and pure, ideologically correct, just and fair, logical or well planned. In fact it is likely to be chaotic, a veritable rollercoaster ride with unforeseen pluses and minuses, moments of grief and despair, and glimmers of hope. But if in the midst of this glorious muddle we can create enough time to manage some sort of landing where a different and sustainable level of population has been achieved, then that is perhaps more than we currently dare dream of. We have no option other than to work night and day to this end.

NOT BY BREAD ALONE

Our understanding of a beautiful and fruitful world started on the back of a large tortoise. It then became a large flat edifice, before morphing into a sphere at the centre of the galaxy, with stars orbiting it, until we understood it was a tiny insignificant planet in an immeasurable universe. It has taken 3000 years of civilisation to get thus far.

Now mankind's thinking has to change again for from an abundance of nature in which to hunt, gather and wander, to small tracts of land round which he could settle, to excessive deforestation and cultivation and industrialised agriculture, he has to embrace the new world of food production at full throttle and kept running on a sustainable system; a great moment of change in a fleeting moment of his history.

To fail will have catastrophic consequences causing huge suffering and possible extinction. The animal kingdom is of course no stranger to extinction. Death is an essential part of the cycle of life and its continuous renewal. Biology is very positive about death and operates on the basis that the individual's death and decay feeds the generations to come. Similarly one species exhausts itself and becomes extinct, and in so doing opens the door to renaissance of life with other and new species.

Biologists talk of the five great episodes of mass extinction in the history of life and we are indisputably now into the sixth with plant and animal life being savaged and exhausted by the change and decline in habitats.

Human inspired technology has not defeated the laws of nature, but rather exploited or defied them for a while, before nature inevitably has its final say. Whether our end is eventually linked to excessive pollution, radiation, starvation, disease or geology is an unknown. Step off the planet and look back and we see the spectacular and rapid rise of one species, the exhaustion of its food, the pollution of its environment and then the collapse of that population. It could be said that man, who mocked the lemming, now runs hell bent towards the cliff edge. There may be a few who survive whilst the earth takes a deep breath before we work once more to repeat the process or maybe not, having learnt our lesson. Put this dialogue into the context of zooming in on the backstreet or arid desert and we see all that extinction already in the eyes of the distressed mother and the dying child.

But it doesn't have to happen like this. I was bought up in the Christian tradition, then, like most, questioned and wrestled with what proved to be a mix of history, superstition, truth, example and faith. As a scientist it was almost expected that I would find the conflict between belief and evidence-based fact to be irreconcilable, but I didn't. When I first looked down a microscope I was amazed at the beauty of the tissue on the slide, and as over the years science has revealed more and more of the secrets of the biosphere and its population and of space with its promise of never ending discovery, I have wallowed in its immensity and poetry. In my short life span science has continued its inexorable progress, often by disproving what had previously been its wisdom. Science therefore reflects man's ability to articulate and like the biology it serves, it is dynamic and its newness never ending. Consequently fresh discovery that makes early science devoid is not a weakness but a strength. However, we have to be on our guard and remember that today's scientific wisdom may be tomorrow's folly; we have not come to the end of science and we are still on the pebbled beach staring out at Newton's ocean of unknowns.

Faith on the other hand has set out some fundamental truths that remain unchanged, for they observe and posses the human spirit. Such truths are the constant in the human equation and when shed leave an irresolvable formula. Of course, religion has been and is still being used to further wrong ends and has often been but a shadow of its founding truths, but, however poorly and inadequately it has been made manifest, it has served humankind well by carrying this particular torch from generation to generation. It is interesting that many of the great scientists and theologians have realised that the human equation is made that bit nearer to complete by the mutual inclusion of faith and science. It is only when the man of faith tries to discredit science or the man of science endeavours to discredit faith, that there is a hollowness to the discourse.

Professor Rees, the Astronomer Royal, has questioned whether man has the capability ever to understand the universe and our place within it, with its apparent eternity of sizes and the irreconcilable conflict between existence in the 'big' state and in the 'small'. Science is governed by laws and therefore is a subjugated discipline. Once it was treated as a philosophy, now it is increasingly described by mathematical formulae. It is on the other hand, love and suffering, that have given us art, music and words that transcend our knowledge and ability to comprehend and explain, but take us to a land beyond, where for the very briefest of brief moments its poetry makes all things clear.

In modern times in the UK to talk of faith has almost become political incorrect and slightly risible. That is not a global phenomenon, but seems to be allied to a country that has become old, less vibrant and perhaps a degree cynical. That for me, living in my Judeo-Christian tradition, is a pity because the wisdom of generations as recorded in the Bible and the teachings of Jesus have a huge resonance that time and time again are proved right in our living and behaviour and could well provide the staff we need as we journey on a difficult path in search of the answer to our question.

When I first learned in biology that ontogeny repeats phylogeny I was amazed and liberated. In biological terms it means that the stages of the evolution of a species are repeated in the development of the individual starting as a single cell and then developing into a mature adult. The amazement was of the wonder at the biology; the liberation was an embryonic understanding that such a concept extended far beyond the science into every part of life.

Re-enter the Garden of Eden, which to some is little more than a piece of fictional irrelevance, whilst to others it is a story retold in many cultures that through its poetry describes our beginnings. The promise was of paradise until Adam, through his breaking of the garden's rules, ate the forbidden fruit and acquired knowledge and for the first time saw and understood his nakedness. In our age of discovery we continue to feast on that fruit and with new knowledge comes new awareness of our nakedness, or lack of wisdom, to deal with the consequences of that knowledge. It is not that seeking new knowledge is unacceptable or bad, for we as humans are cast in that role and it is part of our make up, but the greater the knowledge we possess, the greater the responsibility upon us to somehow use it for good and for the benefit of all; and that requires great wisdom. So our long history as a species is reborn in each generation and the Garden of Eden is just as relevant to us as it was to Adam and his soul mate Eve.

Who then better to learn from than those who have been there before us and who have lived through what we ourselves are now experiencing. Take the story of the flood and how those who had abused the world they lived in were drowned along with the innocent in a great depth of water. That story is still retold almost daily as we witness an ever increasing number of catastrophic floods resulting from climate change, excessive deforestation or extensive urbanisation leaving the water nowhere to go other than up.

And so story after story and saying after saying continue to be played out in the modern world as they have been right down through the ages. Once they were attributed to God's personal intervention as a judgment on mankind, whilst today they are, wherever possible, explained by science. But, however superior or sophisticated we may feel, time and the knowledge it brings has not changed the consequence of actions nor the wisdom needed to avoid or deal with such repeated scenarios.

For a decade I had the privilege of attending some of the great services at Westminster Abbey celebrating Christmas and Easter. My eldest brother was the Dean and he would preach on each such occasion. His sermons were always crammed with ideas and new perspectives which challenged the listener not least in their ability to keep pace. Sermons are rarely remembered, whether delivered in a chapel or Cathedral, but rather act like small grains of sand that are laid down as the words wash over the listener; they accumulate with time, layer upon layer, until they eventually form a rock.

On one occasion at the Midnight Mass on Christmas Eve he talked of the birth of Jesus and the stable scene with its accompanying animals and almost as an aside proffered the thought that maybe amongst those assembled beasts were the four horses of the Apocalypse. Representing pestilence, war, famine and death those fine beasts were the threatened judgment on mankind, but here they were sharing this humble out-building with the child sent to bring peace. It was a little piece of poetry that prompted many questions. For sure those horses are now out of their stable fulfilling their function; nobody would argue other than that. The question though is whether we can reconnect, through the wisdom and understanding of those who have gone before, and once again comprehend the vulnerability of the child that made men stand tall and seek to serve others. However read, there is urgency and immediacy in its telling.

Professor Rees graphically notes that in the term of the sun's existence expressed as a 365 day year commencing 1st January, then the 21st century will represent a quarter of one second sometime in June. He then goes on to ask whether this century will be mankind's last. Such thinking is almost apocalyptic and contains real urgency and immediacy; a modern day version of Revelations. The two messages have a commonality in dealing with the fate of mankind and sharing in their honesty and search for truth. Professor Rees, the scientist, urges us to listen again to the elderly Einstein whilst the Dean, reaching further back, offers us the Christ child.

Whilst Rees toys with the thought that maybe the human mind has not got the capacity to have a full understanding of the universe in which he exists, Einstein, the man he invokes, a Jew educated in a Catholic school, asks a similar question, his being whether man can understand the nature of God. He famously said,

'I'm not an atheist and I don't think I can call myself a pantheist. We are in the position of a little child entering a huge library filled with books in many different languages. The child knows someone must have written those books. It does not know how. The child dimly suspects a mysterious order in the arrangement of the books but doesn't know what it is. That, it seems to me, is the attitude of even the most intelligent human being toward God'.

Perhaps there is not too much distance between Einstein, Rees and the Dean and in fact all are fellow travellers treading the same boards and each giving their particular performance in the same play before us, the audience.

There is therefore no reason for scientific advancement and the story by which we live to battle with each other for now science too must address 'eternity' if man is to survive in a planet of beauty. It is science that can study and direct how we navigate the next clutch of decades and it will require exceptional findings of knowledge if we are to arrive through to the other side with a secure and sustainable existence for our children's children. The deep roots within us, which some call our spiritual being, needs also to awaken to give us the depth of resolve, concern and wisdom to ensure we have the necessary determination and fortitude to make it happen for them. That may be the difference between a story with another chapter or one with an abrupt end.

In the chaos that is already here in part and is sure to come in full to all our lives, perhaps there will be a place for this book, and this particular 'pebble' will be picked up, may be for just a fleeting moment, and considered a little, before being flung, skimming across the onshore waves, into the ocean.